UNCLE SAM'S CABINS

A Visitor's Guide to
Historic U.S. Forest Service Ranger Stations
of the West

Les Joslin

Wilderness Associates
Bend, Oregon

For information address Wilderness Associates,
P.O. Box 5822, Bend, Oregon 97708.

ISBN 0-9647167-1-2

Library of Congress Catalog Card Number: 95-90370

Front cover photograph of Slate Creek Ranger Station,
Nez Perce National Forest, Idaho,
and other photographs not otherwise credited
are by the author.

Maps by the author.

Printed and bound in the United States of America by
Maverick Publications, Inc.
P.O. Box 5007, Bend, Oregon 97708
for
Wilderness Associates
P.O. Box 5822, Bend, Oregon 97708

This book is dedicated to my wife,
Patricia,
truly "the wind beneath my wings" for twenty-five years,
and to
those men and women of the U.S. Forest Service—
archaeologists, historians, forest supervisors, district rangers,
historic preservation teams, volunteers, and others—actively
striving to preserve the heritage
of their unique organization.

*"The past belongs to the future,
but only the present can preserve it."*
—Anonymous

CONTENTS

PREFACE

As close as I can fix it, my interest in historic U.S. Forest Service ranger stations which resulted in this book dates from the afternoon in June 1962 when I arrived at the old Bridgeport Ranger Station to begin my Forest Service seasonal "career" as a fire guard. If that old Toiyabe National Forest ranger station four miles northwest of Bridgeport, California, were still all there, it would be one of the seventy-five profiled in this book. But it isn't. Late in 1962, the one-room, Depression-era Bridgeport Ranger Station office building, replaced by a new structure closer to U.S. Highway 395, was moved to Reese River, Nevada, for use as a guard station. I haven't seen it since.

But, over the years, I have run across many other historic ranger stations—they're historic if built before World War II—on national forests throughout the West. A few years ago, I hit on the idea of doing for historic ranger stations what Ira Spring and By Fish in *Lookouts: Firewatchers of the Cascades and Olympics* (Seattle: The Mountaineers, 1981) and Ray Kresek in *Fire Lookouts of Oregon and Washington* (Fairfield, Washington: Ye Galleon Press, 1985) have done for fire lookouts. Work began in earnest—interrupted only by teaching assignments, wilderness work, fire calls, and family life—in the spring of 1993. Two years, tens of visits and interviews, hundreds of letters and telephone calls, and thousands of miles later, I've come up with *Uncle Sam's Cabins: A Visitor's Guide to Historic U.S. Forest Service Ranger Stations of the West.*

This book tells the stories—or parts of the stories—of seventy-five historic ranger stations found in the Forest Service's seven western regions. It's intended to help national forest visitors appreciate these historic sites and the efforts to restore, preserve, and interpret many of them.

This book has a simple format. After an introductory chapter on forest rangers and ranger stations, the historic ranger stations it profiles appear in chronological order in seven chapters based on

xi

SLATE CREEK
HISTORICAL RANGER STATION

THIS CABIN WAS THE FIRST SLATE CREEK RANGER STATION. UNTIL 1975, THE CABIN STOOD FIVE MILES UP SLATE CREEK WHERE A HOMESTEADER BUILT IT AROUND 1900. THE CABIN BECAME HEADQUARTERS

The Ninemile Remount Depot

The Ninemile Remount was, at one time, the center for U.S. Forest Service packing activities in the Northern Rockies. Completed by the Civilian Conservation Corps (CCC) in 1935, the Remount was the home roost for more than 1500 Rocky Mountain Canaries (also known as mules), as well as prime breeding stock for the Forest Service. Firefighting, trail and fire lookout building, and many other kinds of backcountry work was done with Ninemile packstock wearing US brands on their hips and diamond hitches cinching down their cargo. The pack strings were spurred on by packers who had reputations for never sparing the adjectives when the going got tough.

The Remount's upper hayfield was used to train smokejumpers in the early years of the famous firefighting crews. More often than not,

droning Ford Tri-Motor airplanes had to buzz the field to clear it of grazing livestock before the jumpers could hit the silk.

In 1980, Ninemile was listed on the National Register of Historic Places for its traditional architecture and its role in Forest Service, Civilian Conservation Corps and local history. Ninemile is still a working ranger station and pack depot, the home for a number of USFS activities, combining history and practicality. Mules and horses are still run through the chutes and corrals on their way to perform important work in the backcountry. And if you listen very carefully, you can still hear the canaries sing!

Ninemile Ranger District
Lolo National Forest

OLD ALTA RANGER STATION

This was the first Forest Service Ranger Station. It was built in 1899 by H. C. Tuttle and Than Wilkerson.

In 1904 this site reverted to private ownership under a mining claim. The Lion's Club of Hamilton, Montana, purchased the site in 1941 and donated it to the Forest Service for preservation as a Montana Historic Landmark.

BITTERROOT
National Forest
U. S. DEPARTMENT OF AGRICULTURE

Some historic U.S. Forest Service ranger stations are interpreted as historic sites.

the Forest Service's traditional administrative regions. The word "traditional" reflects the fact that these regions have been around for most of the agency's history, but are subject to consolidation and reorganization. The date assigned each historic ranger station is generally that of construction or original occupation of the oldest existing structure on the site.

Not all these historic Forest Service ranger stations began as ranger stations. A few were built originally as national forest headquarters, from which a forest supervisor oversaw the activities of several ranger districts, and later became district ranger stations. And some that began as ranger stations are now used as guard stations, work centers, visitor centers, or some combination of these purposes. Some are historic sites, pure and simple, and interpreted as such. In this book, any historic Forest Service station is referred to by its ranger station name, not the previous forest headquarters or current guard station name some have had or now have. Such previous and current names are mentioned in each profile.

Some of these historic Forest Service ranger stations are near major population centers, on major highways, and easy to visit. Others are more remote, reached by secondary roads or worse, and more difficult to visit. A couple, in wildernesses, are reached only by trail travel on foot or horseback. Access information is provided for each, but should be used along with current state highway and national forest maps that clearly show the roads and trails to be followed. Mileage distances, sometimes approximations, are derived from odometer readings, road signs, maps, and other sources. Also, travel conditions change over time and vary with seasons, and it's a good idea to check with local authorities for current conditions. Always play it safe, especially when traveling in more remote areas.

I couldn't, and didn't, write this book alone. I had the help of many—including many Forest Service archaeologists and historians—who opened their hearts, minds, and files, and whose words are reflected in these pages. I have chosen to recognize them and other sources in the acknowledgment essay at the back of this book.

Finally, a first effort rarely, if ever, produces the last word on any subject. This book is a first effort, and certainly not the last word, on historic Forest Service ranger stations. I will continue

*Some historic U.S. Forest Service ranger stations
are reached by secondary roads.*

research on historic ranger stations, and will appreciate any additions, corrections, or suggestions that could contribute to a revised, improved, and more comprehensive *Uncle Sam's Cabins.*

Les Joslin
Wilderness Associates
P.O. Box 5822
Bend, Oregon 97708-5822

Seventy-three of the seventy-five historic U.S. Forest Service ranger stations of the West in this book are shown on this map of the national forests of the eleven western states. Another ranger station in Petersburg, Alaska, where ranger boat Chugach *also is moored, are the other two and are shown on the Alaska Region map, page 223. National forest lands are shown in gray on all maps.*

Introduction
FOREST RANGERS AND RANGER STATIONS

Among government employees and edifices, forest rangers and ranger stations are more romantic than most. Both are symbols of the American West's last frontier—a last frontier shared with the cattlemen, sheepmen, lumbermen, miners, homesteaders, and others who used the public forests and rangelands.

The first forest rangers, appointed during the summer of 1898, worked for the General Land Office, the U.S. Department of the Interior agency charged with looking after public lands, including the forest reserves set aside in the West beginning in 1891. With establishment of the Forest Service in the U.S. Department of Agriculture and transfer of forest reserve administration to that new agency in 1905, they became Forest Service rangers. In 1907, the forest reserves these rangers patrolled were renamed national forests.

Gifford Pinchot, the founding Chief of the Forest Service—the title was Chief Forester at the time—who characterized General Land Office administration of the forest reserves as "crooked and incompetent," accepted only the better Interior Department rangers into his new Forest Service. Political appointment of rangers gave way, in Pinchot's outfit, to hiring under U.S. Civil Service Commission rules. This helped Pinchot recruit the caliber of men needed to earn the trust and respect of the American public, and to overcome the animosity inherited from the General Land Office's often inept administration. As Pinchot wrote in his 1946 autobiography, *Breaking New Ground*, the ranger "was the officer with whom the [public] did most of their Forest Reserve business. On how he handled himself and them depended mainly their attitude toward the Reserve and the Service."

1

General Land Office Forest Ranger L.A. Myrick at a Battlement Forest Reserve, Colorado, ranger station about the turn of the century. U.S. Forest Service photo.

The first Forest Service manual, a pocket-sized volume commonly called the *Use Book,* specified that every applicant for a ranger job "must be, first of all, thoroughly sound and able-bodied, capable of enduring hardships and of performing severe labor under trying conditions." In the next sentence, which Pinchot called a "well-deserved . . . slap at the Land Office," the *Use Book* pointed out that "Invalids seeking light out-of-door employment need not apply." According to the *Use Book,* a ranger:

> . . . must be able to take care of himself and his horses in regions remote from settlements and supplies. He must be able to build trails and cabins, ride, pack, and deal tactfully with all classes of people. He must know something about land surveying, estimating and scaling timber, logging, land laws, mining, and livestock business.

That was a tall order for a man also "required to own and maintain his own saddle and pack horses, all for $900 a year."

2

That order was filled by an "examination of applicants along [those] practical lines" taken by men whose motives were as diverse as their backgrounds. As the *Use Book* described the examination:

> . . . actual demonstration, by performance, is required. Experience, not book education, is sought, although ability to make simple maps and write intelligent reports . . . is essential.

After a written test eliminated the illiterates, a practical skills test made sure the Forest Service hired men who could do the job in the woods. Pinchot recalled one such test in *Breaking New Ground:*

> Well I remember a Ranger examination . . . in the Bitterroot Valley of Montana. It required the candidate to prove by doing that he could run compass lines, chop, pack a horse, and find his way by day or night. It also included two other highly practical tests. The first was: "Cook a meal." And the second: "Eat it."

Sometimes, since toting a pistol and rifle often was necessary, these examinations included a marksmanship test.

Whether, as Norman Mclean wrote in his short story, "USFS 1919: The Ranger, The Cook, and a Hole in the Sky," these examinations "picked rangers for the Forest Service by picking the toughest guy in town" or not, early-day forest rangers were rough and ready as well as resourceful, reliable, and responsible. "They had to be pretty tough hombres," explained Betty Goodwin Spencer in her 1956 book, *The Big Blowup,* "if they cared to continue living."

> They were given the glad hand by few and the cold shoulder by many. Lumbermen, railroaders, miners, homesteaders, and ranchers regarded the Far West as pretty much their own private back yard. They didn't take easily any outside interference, and so they fought and bucked the Forest Service with all their ability.

From the beginning, this small and scattered force of early-day rangers patrolled and protected vast reaches of mountainous forests and rangelands under arduous conditions—and, in so doing, they won over the American public. By the time Pinchot finished as Chief Forester in 1910, the Forest Service was well organized, the most violent opposition to its management of the national forests was past, and a second generation of rangers—many of them with college degrees as well as practical skills—was wearing the pine tree shield.

In the man's world of the early Forest Service, it doesn't seem to have occurred to anyone that a woman might do the ranger's job. When, eventually, it did, the old California Region's 1931 *Forest Rangers' Catechism* responded: "Women are not appointed by the Forest Service as members of the field force even if they pass the civil-service examination." In fact, many a ranger's wife worked alongside her husband, often to the extent of being an unofficial assistant ranger. Times have changed, of course, and women are now fully integrated into the outfit.

As rough and ready as these early-day forest rangers—and their wives—were, they still needed places to live and work. They seem to have recognized that fact before their supervisors did. Two of these early rangers, N.W. "Than" Wilkerson and H.C. "Hank" Tuttle, built the first ranger station—a log cabin at Alta on the Bitter Root Forest Reserve in Montana—in 1899 at their own expense. The next year, the federal government funded construction of West Fork Ranger Station on the San Gabriel Forest Reserve in California, and perhaps others. But ranger stations remained the exception rather than the rule for years to come. Like most of his General Land Office predecessors, many an early Forest Service ranger worked from a rented room in town and a tent in the field, and slept out or stayed at the home of a local resident when traveling. The few government-owned buildings—the earliest ranger stations among them—were usually small, poorly designed, and inadequate for conducting daily business.

Chief Forester Pinchot worked to promote a specific image of his new Forest Service in many ways. One of these was establishment of needed ranger stations. As he recalled in *Breaking New Ground:*

> Forest Ranger stations also required attention. National Forests could neither be protected nor utilized without Rangers, and the Rangers could neither live near their work nor maintain their

saddle and pack horses unless Ranger stations, which must include pasture for the stock, were reserved for their use.

Accordingly, orders from Washington withdrew such stations from entry in or near the National Forests all over the West.

With sites for ranger stations "withdrawn from entry" by citizens who otherwise could occupy and gain ownership under any of several public land disposition laws, the Forest Service set about building these ranger stations. In the 1907 edition of the *Use Book,* the Washington Office promised that "all the rangers who serve the year round will be furnished with headquarter cabins" that the Forest Service—in many cases, the rangers themselves—would build "as rapidly as funds will permit. Wherever possible," the *Use Book* specified—in perhaps the first stab at standardizing ranger station architecture to reflect a consistent Forest Service image, these "cabins should be built of logs, with shingle or shake roofs." And, the *Use Book* continued:

> Cabins should be of sufficient size to afford comfortable living accommodations to the family of the ranger stationed in them, and the ranger shall be held responsible for the proper care of the cabin and the ground surrounding it. It is impossible to insist on proper care of camps if the forest officers themselves do not keep their cabins as models of neatness.

A tradition, as well as an image, was being forged. Still more guidance in the 1908 edition of the *Use Book* recognized other needs of pioneer ranger life:

> Rangers' cabins should be located where there is enough agricultural land for a small field and suitable pasture land for a few head of horses and a cow or two, in order to decrease the often excessive expense for vegetables and food.

It went on to say this pasture should be large enough to support the stock needed for the ranger's official duties.

In keeping with this guidance and small budgets, the earliest Forest Service ranger stations normally consisted of a single cabin and, perhaps, a barn along with another outbuilding or two and a

Forest rangers of the newly-formed U.S. Forest Service met at Mesa Lakes Ranger Station, Colorado, in June 1905. They are (left to right) Ranger William R. "Bill" Kreutzer, the first U.S. forest ranger, Ranger (and later forest supervisor) Dave Anderson, Ranger Henry Dingman, Ranger (and later forest supervisor) John W. Lowell, Ranger B.Z. Jay, Ranger Frank Barnes, and Ranger James G. "Jim" Cayton whose Cayton Ranger Station story is told in Chapter Two. Photo courtesy of Helen Spence, Collbran, Colorado, and U.S. Forest Service.

corral. Pinchot's guidance was clear enough, but various interpretations of new regulations by field personnel working in diverse environments thousands of miles from the Washington Office often resulted in a lack of uniformity in field operations—including planning and construction of ranger stations. From the beginning, ranger stations varied from one part of the West to another. In the arid Southwest, for example, ranger station builders often used such local materials as adobe brick and stone. And, of course, some ranger stations were previously existing buildings—some of them abandoned miners' cabins and cowboys' line shacks—pressed into service. Nevertheless, as specified in the *Use Book,* early ranger station architecture was epitomized by the simple log cabin. As time passed, other factors affected how and where the Forest Service built ranger stations. These included the administrative decentralization of the agency, changes in national forest management, advances in transportation and communications, and access to construction resources.

In 1908, three years after it was established, the Forest Service decentralized. Six district offices under district foresters—renamed regional offices and regional foresters in 1930—at Missoula, Denver, Albuquerque, Ogden, San Francisco, and Portland were set up between the Washington Office and the national forests. Also, in 1908, the national forests were subdivided into ranger districts, still the basic administrative unit of the National Forest System. District rangers were soon assigned, and ranger stations were their headquarters. These rangers reported to forest supervisors, who in turn reported to the district—later regional—foresters, who answered to the Chief Forester in Washington, D.C.

Most ranger district staffs remained small during the Forest Service's first few decades. A ranger, who did most of his district's work himself, usually was aided by a clerk and perhaps a few seasonal assistants such as fire guards. Ranger stations were small, and other even smaller facilities—later called guard stations—were dispersed within ranger districts to provide shelter for the ranger and his assistants when in the field. Many rangers of this era didn't occupy their ranger stations year round, but were detailed to their supervisor's office or even to the district forester's office during the winter. But their small ranger stations, where they worked with the public during the summer field season, became Forest Service symbols.

At the same time, in many parts of the West, the log cabin ranger stations specified in the *Use Book* were quickly being replaced by rustic, wood-frame structures. The image of the Forest Service was evolving, and crude log cabin ranger stations didn't project an image of cleanliness, efficiency, and dedication to the public it served—at least not in the minds of many agency leaders. The log cabin did persist in the Rocky Mountain national forests, especially in the Rocky Mountain Region itself where log construction continued to be popular through the Great Depression era of the 1930s.

The old California District, now the Pacific Southwest Region, was among the first—if not the first—to attempt standardization of ranger station architecture. On May 1, 1917, District Forester Coert DuBois in San Francisco issued an improvement circular that set forth "designs for buildings . . . generally." It called for standard wood-frame construction of large buildings, and smaller log buildings. These buildings were small and inexpensive to erect; they cost as little as $112 in labor and materials. The designs reflected the Craftsman style of the era, and were drawn with an eye to more than strictly functional requirements. They were to promote that Forest Service image. DuBois's plans were adopted by many California national forests, but varying grades of lumber, accessibility, costs, and individual preferences resulted in ranger station buildings that often differed from the original plans.

A second era of Forest Service ranger station construction—perhaps one of transition from the rustic log cabin of the early 1900s to the more modern compound built during the Great Depression—seems to have occupied the 1920s. After the disastrous 1910 and 1919 fire seasons in Idaho and western Montana, Congress mandated increased operations to contain wildfires. Also, the Forest Service had begun a transition from "custodial superintendence" to "active management" of national forest resources. This increased the work force, and required larger facilities. As a result, to the extent limited budgets allowed, existing ranger stations were either expanded, or moved and expanded.

Standardization, at least within regions, remained an issue. The old California Region, for example, was quite specific about the appearance of its ranger stations. "A ranger station is painted gray with white trim, has a green roof, and is marked by a sign bearing the name of the station," its 1931 *Forest Rangers' Catechism* stated. "The American flag is always flown from a flagpole in front of the station when the ranger is at home." Within a few years of that categorical pronouncement, however, the region's ranger station paint scheme changed from the French-battleship gray paint that the June 16, 1933, issue of *California Ranger* said had "depressed the morale of the rangers for fifteen years" to a tobacco brown. Only the flag, it seems, has remained a constant.

Centennial Ranger Station, Medicine Bow National Forest, Wyoming, in 1940, was a typical Depression-era ranger station compound. U.S. Forest Service photo.

A third era of ranger station construction—the last "historic" era—was ushered in by the Great Depression. A pivotal time in American economic, social, and political history, the Depression marked also a turning point in the Forest Service's mission—and the ranger stations it needed to do its work. As part of his New Deal to combat unemployment, President Franklin D. Roosevelt created the Civilian Conservation Corps (CCC) to employ young men in conservation work. Operated by the War Department, national forest CCC camps focused on improving or replacing old, substandard facilities, including ranger stations. Among other sources of labor for such projects were older local experienced men, called LEMs, also in need of employment. Until the Depression, the Forest Service budget for constructing ranger stations and other facilities was quite limited. Since building cost limitations did not apply to Depression-era projects, ranger station construction boomed. Ranger stations built between 1933 and 1941, clearly reflecting the Forest Service's expanded mission, were seldom small cabins. A range of building types—offices, residences, barracks, warehouses, garages, and shops—arranged in efficient compounds symbolized the change from custodianship to active management of the National Forest System.

Perhaps because the Washington Office recognized that the Forest Service operated throughout a West in which no one style would be appropriate to all regions, the regional offices were allowed to design standard ranger station buildings appropriate to their regions' natural environments. Regions took advantage of native or pioneer building styles and materials to create "accessories of nature." Thus, in the Northern and Rocky Mountain regions, the Rocky Mountain Cabin style of log construction continued. Pueblo style structures of adobe and stucco appeared in the arid reaches of both the Rocky Mountain and Southwestern regions. But, perhaps, the most distinctive styles developed in the California and Pacific Northwest regions.

With a variety of New Deal emergency relief work programs providing funds and labor for numerous projects, Regional Forester S.B. Show, who succeeded Coert DuBois in San Francisco, hired a staff of architects to plan for "a renaissance in Forest Service ranger station architecture." Among those architects, Norman Blanchard and Edward J. Maher of the San Francisco firm Blanchard, Maher, Spencer and Hall were in charge of designing administrative structures. According to the June 16, 1933, issue of *California Ranger*, these architects would help replace an "amazing variety of stations built (or acquired) in this free for all period" since 1900 that included "trappers' cabins, miners' shacks, cow-punchers' bunkhouses, ranchers' homes, and the bungle-ohs of the southern Californian from Iowa" with structures that would "combine the last word in art, comfort, and utility," bringing a consistent image to California Region ranger stations. The publication went on to predict:

> Not only will the lines of our ranger stations be revamped, but the color scheme will be improved. The green roof will be retained, but the French-battleship gray paint . . . will be changed to a brown stain to blend appropriately with the colors of the forest.

These architects' Mother Lode style was influenced by both the Craftsman style of the 1920s and the California Ranch style of the 1930s. Their "ready cut" designs, built throughout the state by Depression-era labor, remain major contributions to Forest Service architecture. During the 1930s, when over 1,200 new buildings—including those of many ranger station compounds—were built in the California Region, many older buildings were removed, burned, or sold. As a result, pre-1930s ranger stations are relatively rare in the Pacific Southwest Region. Yet, even with this effort, complete standardization eluded the region. This lack of uniformity was reflected in the fictitious Ponderosa National Forest supervisor's remarks to one of his district rangers in George R. Stewart's classic 1948 novel, *Fire*, that he appreciated seeing a ranger station "in white and green occasionally instead of all that tobacco-brown we go in for so much."

The distinctive style that developed in the Pacific Northwest Region, a Cascadian Rustic style that made lavish use of sturdy timber, was intended to reflect the Forest Service's mission of fostering the production and use of wood in timber-rich Oregon and Washington. At the same time, this labor-intensive style of ranger station construction provided a lot of work for CCC and other work relief program enrollees. After the Depression, labor-intensive projects became uneconomical, and such buildings were never built again.

Yet, not all these standard plans were rustic. In the Northern Region, for example, Cape Cod style ranger station compounds were built. Although deviations from these standard plans required regional office approval, many deviations occurred throughout the regions. Many changes to standard plans were made by CCC construction crews, most likely for very practical reasons, and remain as evidence of their ingenuity and resourcefulness. Still other deviations occurred for other reasons. In the Pacific Northwest Region, for example, a unique Colonial Revival style ranger's residence was built from a *Ladies Home Journal* plan at the ranger's wife's request. Throughout the West, many of these Depression-era ranger station structures share a distinctive mark: the "pine tree logo" of the Forest Service and the CCC found in many shapes, sizes, and forms on the buildings of the period.

Down through the years, the fortunes of individual Forest Service ranger stations built before World War II have waxed and waned as national forest management objectives evolved, and transportation, communications, and technology improved. As the Forest Service transitioned from custodianship to active management of national forests, many large ranger districts were "split" into smaller districts to facilitate a new management emphasis on harvesting timber and building roads. Ranger stations were built for these districts. Later, improved transportation and communications made larger ranger districts more practical, and many smaller districts were "lumped" together or consolidated. Ranger district staffs and facilities—including ranger stations—increased in size and sophistication. As a result, many ranger stations built before or during the Depression became guard stations or work centers, and some ranger and guard stations that were no longer needed were either abandoned, dismantled, or destroyed.

Today, a diverse collection of historic U.S. Forest Service ranger stations remains in the national forests of the West. Many, but not all, of these visible and visitable symbols of Forest Service history have been added to the National Register of Historic Places. Some remain in use as ranger and guard stations. Others are preserved and interpreted by a Forest Service—often with the needed help of its Passport in Time program volunteers or other volunteers—that values its unique place in America's history. Still others are abandoned, their futures uncertain. A visit to any is a step back in time that often evokes the "one-man" ranger districts of the not-too-distant past when, as fictitious Ranger Varick McCaskill in Ivan Doig's 1984 novel *English Creek* put it, anyone who was "going to get by in the Forest Service" had "better be able to fix anything but the break of day."

Chapter One
NORTHERN REGION

The national forests of the Northern Region, which the Forest Service also calls Region 1, stretch from the prairies and badlands of the Dakotas, through eastern Montana's rolling hills and isolated ponderosa pine woodlands, to the rugged peaks and timbered canyons of western Montana and northern Idaho. These diverse twenty-five million acres of the National Forest System, contained in twenty-three national forests in 1913, have been reorganized over the years into thirteen larger administrative units. One of these, the Idaho Panhandle National Forests, includes the Coeur d'Alene, Kaniksu, and St. Joe national forests. So, in fact, there are fifteen national forests in Region 1. They are managed by a regional forester in Missoula, thirteen forest supervisors, and about sixty-five district rangers.

Twelve historic U.S. Forest Service ranger stations of the Northern Region (Region 1) are in northern Idaho and western Montana.

Region 1 is, in many ways, the "culture hearth" of the Forest Service. Many of its influential leaders—regional foresters like William B. Greeley (who later became Chief Forester) and legendary Major Evan W. Kelley, role models like Elers Koch, and heroes like Ranger Edward C. Pulaski—made their marks there. Many of its pivotal events—the 1910 forest fires, the development of smokejumping—occurred there. America's first ranger station, Alta Ranger Station on the Bitter Root Forest Reserve, was built there in 1899 when much of the region remained a frontier wilderness. About five million acres of the region remain congressionally-designated wilderness. And, among the dozen Region 1 historic ranger stations featured in this chapter is Moose Creek Ranger Station on the Nez Perce National Forest, the Forest Service's last operating wilderness ranger station.

ALTA RANGER STATION
Bitterroot National Forest, Montana
(1899)

Alta Ranger Station, a small log cabin deep in the Bitterroot National Forest, is generally recognized as the first U.S. Forest Service ranger station—even though it predates creation of the Forest Service by almost six years. Built in the spring of 1899 by newly-appointed U.S. Department of the Interior forest rangers N.W. "Than" Wilkerson and H.C. "Hank" Tuttle, this historic one-time ranger station gives visitors a good feel for the rigors and hardships faced by the earliest rangers.

Rangers Wilkerson and Tuttle were among the first men in the United States appointed as forest rangers on the forest reserves then under Department of the Interior custodianship. In June, 1899, these pioneer rangers were posted along the West Fork of the Bitterroot River to patrol some 300,000 acres of rugged Montana backcountry in the newly-created Bitter Root Forest Reserve. Their main jobs were to prevent timber theft and fight forest fires. They needed shelter and, out of necessity, built the rustic, sod-roofed, one-room Alta Ranger Station cabin. It took Rangers Wilkerson and

14

General Land Office forest rangers N.W. "Than" Wilkerson and H.C. "Hank" Tuttle raised the American flag at Alta Ranger Station on July 4, 1899. U.S. Forest Service photo by N.W. Wilkerson.

Tuttle, who borrowed a workhorse to skid logs to the site, fifteen days to build the cabin. As Ranger Wilkerson wrote in his diary:

> Ranger Hank Tuttle and myself constructed the Alta Cabin on our own initiative. . . . we went down into our own pockets for every cent. . . . Uncle Sam wasn't putting out a good cent for a pair of hill-billy rangers to live while they took care of his woods.

And so the two rangers, whose pay was fifty dollars per month, bought the hardware and one glass window for the log cabin with their own funds. But, as Michael Frome put it in *Whose Woods These Are*, "when they erected a flagpole and hoisted a flag aloft, this *was* a ranger station." On July 4, 1899, the two rangers dressed up in their "Sunday best" and had their picture taken with the new cabin and the flag they also had purchased with their own money. Alta's flying of Old Glory that day began the custom of flying the flag at ranger stations.

At the time Alta Ranger Station was built, the only other habitation on the West Fork of the Bitterroot was a lone prospector's cabin. As mining intensified in the area, the community grew. An assay office, post office, store, inn, and saloon were built during the opening years of the twentieth century and, in 1909, a telephone line connected Alta to the outside world. For some reason, the ranger station site was not withdrawn from entry as a homestead or mining claim under the public land and mining laws, and the land was patented as a mining claim in 1904. The cabin was sold to a private owner and moved to Darby that same year. As a result, the Forest Service—which took over the forest reserves in 1905 and renamed them national forests in 1907—withdrew twenty-seven acres adjoining the original site and built another ranger station in 1908. It no longer exists.

Rangers Wilkerson and Tuttle couldn't have known in 1899 that their hastily-built 12-foot by 16-foot log cabin would become an historic site—listed on the National Register of Historic Places seventy-six years later as the nation's first ranger station. It's not much to look at, but its significance has long been recognized. Back in 1941, the Hamilton, Montana, Lions Club purchased the site of the original Alta Ranger Station, moved the restored cabin to the site, and deeded both to the Forest Service for preservation and

Historic Alta Ranger Station, Bitterroot National Forest, Montana.

interpretation as an historic landmark. But the cabin again fell into disuse and neglect, and was in poor shape when Peyton Moncure wrote about "The First Ranger Station" in the October 4, 1959, issue of *The Salt Lake Tribune Home Magazine*:

> Only the pack rats scurry among the old newspapers and magazines that litter the floor now, and a few old window panes are broken out, but the day may come when this old landmark will be something of an official monument to pioneer foresters who had such a personal interest in protecting the nation's forests that they were willing to sacrifice much of their own time, comfort, and meager funds.

That day came. Renewed interest in historic building preservation in the 1960s led to a second restoration. Historic Alta Ranger Station, which now receives regular Forest Service and private financial support, remains to tell its tale.

ACCESS: Alta Ranger Station is located about 30 miles south of Darby, Montana, just off Montana Highway 473 which becomes Ravalli County Road 104. Turn off U.S. Highway 93 about four miles south of Darby, and follow this paved highway, which passes the West Fork Ranger Station, for 20 miles. Continue southward on the intermittently-paved road for another 10 miles, skirting the east side of Painted Rocks Lake (a reservoir), to Alta Campground. Just past the campground, turn left onto Forest Road 13 and drive about 1/4 mile through a rustic farmstead to a cabin on the left which an interpretive sign identifies as "Old Alta Ranger Station."

BULL RIVER RANGER STATION
Kootenai National Forest, Montana
(1907)

Bull River Ranger Station was built in the earliest days of the Forest Service by Granville "Granny" Gordon, the first ranger of the Noxon Ranger District of the old Cabinet National Forest—now part of the Kootenai National Forest. Ranger Gordon, a cowboy, was a friend of President Theodore Roosevelt.

When the search for a Noxon District summer ranger station site focused on Marion Cotton's pasture, uninhabited and claimed only by squatter's rights, Cotton protested in an emotional letter to Chief Forester Gifford Pinchot. But the Forest Service took the land, and paid Cotton a hundred dollars to cover the cost of his improvements.

Ranger Gordon, given a free hand to design the station, drafted plans for a two-story cabin with a kitchen, a sitting room, three bedrooms, a closet, and two porches. Construction began in January, 1907, and was completed in 1908. The ranger, his wife Pauline—a one-time cook for Buffalo Bill Cody's Wild West Show, and daughters Grace, Stella, and Blanche moved in that year.

Bull River Ranger Station looked like a typical Montana mountain homestead complete with an orchard, a garden, and outbuildings. In addition to serving as an administrative site, the ranger station became the social center of the community as Mrs. Gordon's

Coyote Kid (left) and Ranger "Granny" Gordon's family at Bull River Ranger Station on the old Cabinet National Forest, Montana, in 1908. U.S. Forest Service photo.

gracious hospitality promoted good will between the Forest Service and the local residents. The station survived the devastating 1910 forest fires, flooding, and vandalism before it was retired in 1970.

A dedicated group of local citizens has joined with the Forest Service to preserve this historic ranger station. In the summer of 1989, Kootenai National Forest employees, Cabinet Wilderness Historical Society members, and other volunteers donated hundreds of hours of labor—weathering frequent rainstorms and blistering heat while coping with enraged pack rats and annoyed bats—to begin preservation work. Between 1990 and 1995, the Northern Region's historic preservation team of skilled craftsmen trained in historic building preservation techniques completed stabilization of the station. An interpretative exhibit is on the site.

ACCESS: *Historic Bull River Ranger Station is located in northwestern Montana, north of the town of Noxon. Turn off Montana Highway 200 onto Montana Highway 56 about 6 miles northwest of Noxon or about 12 miles southeast of the Idaho border, and follow*

19

Montana Highway 56 up the Bull River for about 8 miles to Forest Road 407, not far past the first crossing of Bull River, where a sign points toward the historic ranger station. At the sign, turn right onto Forest Road 407, and follow it for about 3 miles to a right fork, also marked by an historic ranger station sign, onto Forest Road 2278. Continue on Forest Road 2278, cross the first bridge over the East Fork of the Bull River, and proceed to historic Bull River Ranger Station.

AVERY RANGER STATION
St. Joe National Forest, Idaho
(1908)

The old Avery Ranger Station—now called Avery Work Center, and not to be confused with the current Avery Ranger Station at Hoyt Flat—began as a Forest Service tent camp in 1907. It has been in continuous use as a St. Joe National Forest administrative site ever since. But it almost didn't survive its second year.

Ranger Ralph Debitt, the first Forest Service ranger in the northern Idaho town of Avery, completed a log cabin at the tent camp site in October, 1908. This cabin is the oldest structure at the site. Built for $528.38 by a Forest Service carpenter team led by Hans Nelson, it served as the ranger station office until 1967 when the Avery Ranger District's headquarters were moved to Hoyt Flat. Today this handsome log building serves as a Forest Service employee residence.

Less than two years after this cabin's completion, the fledgling Avery Ranger Station almost fell victim to the 1910 forest fires that devastated much of Idaho and western Montana. Avery Ranger Station was a coordination and supply center for hundreds of firefighters attempting to suppress those fires. Then came the famous "big blowup" of August 20 and 21. Twenty-eight of Ranger Debitt's firefighters failed to heed orders to return to Avery and were burned to death three miles from town. Avery was soon surrounded by fire and seemed doomed. Women and children crowded onto trains and were carried toward Missoula and safety,

Historic Avery Ranger Station, St. Joe National Forest, Idaho.

while Ranger Thaddeus A. Roe—his arms and legs already burned so badly he was scarred for life—and seven or eight men stayed to try to save the town and the ranger station. In a desperate effort, they set "backfires" to rob fuel from the advancing flames. The backfires worked and, unlike some other towns that burned to the ground, Avery—including its ranger station—was saved. At the same time, about twenty miles north near Wallace, Idaho, Ranger Edward C. "Big Ed" Pulaski, saved most of a crew of forty-five firefighters when he herded them into an abandoned mine tunnel, held them there as the fire passed over, and became a Forest Service legend.

After the 1910 fires, Avery Ranger Station grew. The site was improved and, by 1911, a two-story house, spring house, wood-shed, and a small barn had been added. A two-room bunkhouse was completed in 1912. Two log cabins, still in use as Forest Service residences, were built in 1922 and 1923. A large log bunkhouse, also still in use, was built in 1928. Other buildings, including a cookhouse that "fed many hungry mouths before it closed down in 1960 and was torn down in 1970" and a large barn that supported over a hundred pack mules but burned in 1965, have come and gone. The historic 1908 office building, almost sold as excess when the current Avery Ranger Station was built, was saved by local citizen action and added to the National Register of Historic Places in 1973.

Avery, by the way, was a major division point for the Chicago, Milwaukee, and Puget Sound Railway, and was named for Avery Rockefeller, a son of wealthy William Rockefeller who owned stock in the railroad. A successor to the short-lived Pinchot Ranger Station three miles down the St. Joe River, Avery Ranger Station was first called North Fork Ranger Station. But as the town of Avery grew, local folks referred to the "Avery Ranger Station" and, over the years, the name became official.

ACCESS: The old Avery Ranger Station is located in the town of Avery, 47 miles up the paved St. Joe River Road from St. Maries and several miles east of the current Avery Ranger Station. In Avery, turn off the road at the old depot building—now a community center—and follow the short dirt drive up the hill to the Avery Work Center. An interpretive sign faces the 1908 ranger station cabin. Please respect the privacy of its residents and of other Forest Service employees who live in the other historic dwellings and the new log bunkhouse built in 1981.

JUDITH RIVER RANGER STATION
Lewis and Clark National Forest, Montana
(1908)

Nestled in a sheltered meadow where the Middle Fork of the Judith River bends through the timbered foothills of the Rocky Mountains' Little Belt Range, historic Judith River Ranger Station—now Judith Guard Station—is one of central Montana's oldest ranger stations. The station's main log buildings, a cabin and a barn, were built by Ranger Thomas G. Myers, assigned there in 1906 by the Lewis and Clark Forest Reserve. Ranger Myers' 1944 telling

Judith Guard Station, Lewis and Clark National Forest, Montana, in 1992. U.S. Forest Service photo.

of the story recalls the early days of the Forest Service when rangers often built their own stations.

In 1908 I received my authorization for $450 for building the dwelling at the Judith Station, blueprints and all made by an engineer by the name of Work. The building was 24x24 with hip roof. The lumber was bought in Great Falls, shipped to Bench-land, and I had to have it hauled from there by team 26 miles to the station. It was about 13 miles to where the logs were cut, at a cost of $1.25 per log, which was not bad.

What got me down was when I started studying the blueprint and found it called for the hip rafters to be 24 feet long. I called the office and told them I thought the hip rafters should be cut 20 feet long, but I got the "No, don't do it" answer (and was told) to follow the blueprint; so I used the full 24 feet for pattern, but when I put two of them up I saw the roof with that pitch would split a raindrop. So I cut off two feet and tried that, but they were still too steep, so I cut off two feet more and it was still quite steep, but I let it go at that.

23

I did not have a level so I wrote to the office for authority to buy one, but got word back no money to purchase level, so I bought myself one.

When I got the walls up and roof on, all I had was one big room downstairs and plenty of room upstairs but no floor. I had a good neighbor at the American Sapphire Mine who said I could use their team and sled to haul logs to their mill, so I went over and cut about 3 [thousand] feet of logs and sawed them into boards and 2x4 and put in partitions and a rough floor upstairs. I still had lumber left, so I put a porch across the front and one half-way across the rear of the house; but no shingles for the roof. The Supervisor came into the station, and in looking things over asked me where I got the lumber. I just said the porches looked awful without shingles, and by gosh I got them. When the fiscal year ended the office had to turn back $3300 of unspent funds. Was I mad? No. I finally got the dwelling in fair condition.

Ranger Myers built the log horse barn in 1909, and is thought to have built the wood-frame garage—probably to house the district's first pickup truck—in 1925. The Myers family lived and worked at the station until 1931.

After 1931, the station was occupied seasonally as a guard station until 1981. Except for modifications limited to roof reshingling and chimney rebuilding after a chimney fire in the 1930s, few changes were made at the Judith River Ranger Station or its environment. The round corral just northeast of the barn was built in the 1960s by then District Ranger Doc Cornell. A modern outhouse northwest of the cabin and modern picnic tables and grills south of the cabin, down by the river, have been added for visitor comfort. The old station deteriorated over time until, in April, 1992, the Northern Region's new historic preservation team and volunteers restored the cabin. Work included rebuilding the foundation and front porch, replacing sill logs, and reshingling the roof. Future improvement plans included window reglazing, wall log face splicing, and restoring and furnishing the interior to the 1920s era. Ultimately, historic Judith River Ranger Station will be developed as an interpretive site and also used as an administrative site.

Judith River Ranger Station, Lewis and Clark National Forest, Montana, about 1915 (above), and Ranger Thomas Guy Myers and family on the station's front porch (below). Photos from the collection of Marian Jeanne Setter, courtesy of U.S. Forest Service.

Northern Region Historic Preservation Team and Lewis and Clark National Forest employees and volunteers replaced Judith Guard Station's shingle roof in 1992. U.S. Forest Service photo.

ACCESS: *Judith Guard Station is located on the Middle Fork of the Judith River in central Montana. The nearest town is Utica, on Montana Highway 239 between Hobson and Stanford. Services in Utica are limited to a combination bar-cafe-convenience store; there is no gas station. To reach Judith Guard Station from Utica, drive southwest on the main gravel road, Forest Road 487, for 12 miles to the Y-junction with Forest Road 825. Keep right at this junction, and follow Forest Road 825 for about 1 1/2 miles to the station. The site includes the historic ranger station complex, a picnic area, and a handicapped-accessible outhouse.*

Slate Creek Ranger Station, Nez Perce National Forest, Idaho, in its original location, in 1923. U.S. Forest Service photo.

SLATE CREEK RANGER STATION
Nez Perce National Forest, Idaho
(1909)

Frank Hartman, second ranger of the Slate Creek Ranger District from 1911 to 1915, lived with his family in this two-story log cabin five miles up Slate Creek from its present location at today's Slate Creek Ranger Station. The cabin, built in 1909, served as a district ranger headquarters from August of that year until April 1917 when the district office was moved to Adams Ranger Station, and as a guard station used by fire crews, trail crews, and road crews until 1959. It was moved to its successor Slate Creek Ranger Station in 1975.

Today, the completely restored and refurnished Slate Creek Historical Ranger Station (shown on the cover) offers visitors a glimpse of early Forest Service family life. As the story is told in the cabin, Ranger Hartman spent many days away from his family

27

Slate Creek Historical Ranger Station, Nez Perce National Forest, Idaho (above). Completely restored and refurnished, Slate Creek Historical Ranger Station offers visitors a glimpse of early-day Forest Service family life. Ranger Hartman's "office" occupied one corner of the cabin (below).

28

as he patrolled his district on horseback. His daughter, Frankie, was born on such a day. Since the doctor could not arrive in time for Frankie's birth, a neighbor down the creek helped deliver the child.

The Hartmans lived a frontier life. They tended a garden and orchards. They purchased groceries and other supplies fourteen miles away in White Bird, and had them delivered to the settlement of Freedom—today's Slate Creek. The family hung its meat from the cabin's ceiling joists, and kept other perishables in a root cellar. A nearby creek supplied water. In a shed between the cabin and the barn, Ranger Hartman used a forge and anvil to fashion horseshoes for his saddle and pack stock. The children spent most of their time helping their mother with chores, but occasionally were allowed to ride with their father on patrols. A corner of the cabin served as the ranger's office, where he spent the cold winter months catching up on paperwork at his desk.

Eventually, the Hartmans moved into Freedom so the children could go to school. Virgil "Slick" Hartman, one of the younger Hartman children, worked for the Forest Service at Slate Creek Ranger Station until his retirement in 1985—long enough to see his family's one-time home become a museum.

The cabin was dismantled by the Nez Perce National Forest Hotshots, an elite fire crew—which found and removed fourteen rattlesnakes from the log structure—and was moved to its current location in 1975. After it was reassembled, the original Slate Creek Ranger Station was given a new floor and shake roof. Many of the furnishings and artifacts that contribute to the cabin's early ranger station appearance were donated by former Forest Service employees.

ACCESS: Slate Creek Historical Ranger Station is located at Slate Creek Ranger Station on U.S. Highway 95 about 20 miles south of Grangeville. Forest Service employees at the ranger station are pleased to provide access to the cabin.

MOOSE CREEK RANGER STATION
Nez Perce National Forest, Idaho
(1921)

The past and the future share the present at Moose Creek Ranger Station, the Forest Service's last operating backcountry ranger station and, until 1995, headquarters of the National Forest System's only entirely wilderness ranger district. At 559,920 acres, the Moose Creek Ranger District comprised the Nez Perce National Forest portion of the 1,337,681-acre Selway-Bitterroot Wilderness of Idaho and Montana. Every year, from April to November, the district ranger and his staff at this historic log cabin community meet the challenge of accommodating a variety of human uses while preserving wilderness values as required by the Wilderness Act of 1964.

In the early days, rangers on the old Selway National Forest worked the Moose Creek country—transferred to the Bitterroot National Forest in 1934 and to the Nez Perce National Forest in 1956—from an abandoned homestead called Three Forks Ranger Station. Jack Parsell, Moose Creek ranger from the district's establishment in 1920 to 1922 and again from 1945 to 1955, moved his new district's headquarters and his new bride to the current Moose Creek Ranger Station site in 1921. "I set up a tent headquarters . . . and built the building that is now used for the cookhouse and administration building," he recalled in 1957. Bert Cramer, who spent the summer of 1921 working for Ranger Parsell, helped build what he called "the Parsell honeymoon log cabin" and remembered that "Jack broke his new bride in right that summer, cooking for the gang, tending telephone, milking the cow, etc., as all good rangers' wives did in the old days."

This first Moose Creek Ranger Station building, a 30-foot by 40-foot cabin that doubled as an office and cookhouse, was similar to other Forest Service structures of the time. It was built primarily of native materials—logs, hand-split cedar shakes, and native granite. Anything not locally available was packed in by animal, as were most of the interior furnishings and personal items. And, as at most early ranger stations, a barn was built. Two residences, a warehouse, and other log buildings were added in the early 1930s.

*Moose Creek Ranger Station, Selway-Bitterroot Wilderness,
Nez Perce National Forest, Idaho. U.S. Forest Service photo.*

By the time Ranger Parsell returned to the Moose Creek District in
1945, the station had grown considerably under the leadership of
the four rangers who had succeeded him.

And it had made Forest Service history. An airstrip, constructed
at Moose Creek Ranger Station "with muscle-power and mules" in
1931, became a development center for backcountry aviation. Not
long after the Forest Service began experimenting with parachuting
firefighters and supplies to fires in remote areas, one of the first
smokejumper bases was established at Moose Creek in 1940. Flying
from the station that summer, pioneer smokejumpers Rufus Robinson and Earl Cooley made the first operational jump on a fire at
Marten Creek. Smokejumping, of course, continues. Another kind
of Forest Service history, wilderness management history, continues to be made at Moose Creek Ranger Station.

Ranger Parsell would feel right at home at today's Moose Creek
Ranger Station, managed by the Forest Service to retain the
appearance of a working backcountry ranger station of the 1920s

Moose Creek Ranger Station, Nez Perce National Forest, Idaho, in 1922 (above) and 1938 (below). U.S. Forest Service photos.

and 1930s. Both living and working conditions reflect his era as visitors experience a nearly forgotten lifestyle dictated by the "minimum tool" concept so important to wilderness management. At the station, dirty clothes are washed by hand before being squeezed through a hand-cranked wringer. Shower water is heated by a wood-fired stove. Firewood is bucked by crosscut saw or hauled or skidded with mule power. And so on.

At work, "primitive skills" are the rule. The ranger station and its airfield—administrative sites and pre-existing airfields are "nonconforming uses" allowed by the Wilderness Act when necessary— are maintained without resort to power tools and equipment, and motor vehicles are completely absent. Two mules, for example, have replaced the tractor once used to maintain the grass runways. Two fire lookouts on the district, employed to minimize use of fire detection aircraft over the wilderness, are packed in by mule string at the start of fire season. Trail crews combine the latest in high-tech backpacking gear with traditional tools of the trade to build and maintain trails. Necessary trail structures are built of native rock and wood to protect soil and water resources and minimize visual impacts.

Although employing the minimum tool principle and using primitive skills carries a cost, this approach is vital to effective wilderness management—including demonstrating and furthering wilderness ethics—at which Moose Creek Ranger District is a nationally-recognized leader. A commitment to reviving the tools and skills necessary to lead the Forest Service wilderness management program toward a mature future is at the heart of the unique Moose Creek Ranger Station way of life.

ACCESS: Moose Creek Ranger Station may be reached by trail or air. The shortest trail route leads 25 miles from Selway Falls, about 21 miles up the Selway River from Lowell, Idaho. Forest Road 223, which passes historic Fenn Ranger Station (pages 52 to 55) to reach this trailhead, is usually open by April. Before starting up this moderate and pleasant trail into the Selway-Bitterroot Wilderness, visit the Selway Falls wilderness ranger cabin for current trail condition information. The public airstrip at Moose Creek Ranger Station provides access to the station and to the heart of the

Selway-Bitterroot Wilderness. It is about 60 air miles from both Grangeville, Idaho, and Missoula, Montana. The airstrip, at 2,400 feet above sea level, consists of two grass runways, one 2,100 feet long and the other 4,100 feet long. There is no unicom system. The Moose Creek ranger's office, upstairs in the Grangeville Post Office building until, as a result of a 1995 ranger district consolidation, it is moved to historic Fenn Ranger Station, or local airports can provide current flying condition information. As for frequencies, Moose Creek Ranger District advises that 122.75 is used for pilot-to-pilot contact, and 122.9 is used for landing at Moose Creek Ranger Station and for announcing intentions throughout the Sel-way-Bitterroot Wilderness. For safety, use both. Commercial flights to Moose Creek Ranger Station may be arranged with commercial pilots in Grangeville, Missoula, and other surrounding towns.

SPOTTED BEAR RANGER STATION
Flathead National Forest, Montana
(1924)

There's been a Spotted Bear Ranger Station near the confluence of the Spotted Bear River and the South Fork of the Flathead River since Ranger John Sullivan built the first cabin to bear that name in 1906. That original station, which no longer exists, was replaced between 1924 and 1934 by three log structures built about a mile and a half to the west by Victor "Big Vic" Holmlund of the Forest Service. These buildings now comprise the Spotted Bear Ranger Station Historic District. This historic ranger station is adjacent to the new Spotted Bear Ranger Station that was dedicated in July 1985 by the cutting of a log with a "misery whip" instead of a ribbon with scissors.

Ranger Sullivan's first Spotted Bear Ranger Station was built one year after the Forest Service was created in 1905 and two years before the Flathead National Forest was carved out of the old Lewis and Clark Forest Reserve in 1908. As headquarters of one of the eight ranger districts formed on the Flathead National Forest in 1909, remote Spotted Bear Ranger Station was the center of

Historic ranger's residence, office, and warehouse at Spotted Bear Ranger Station, Flathead National Forest, Montana. U.S. Forest Service photo.

operations for over 800,000 acres of national forest backcountry. By 1910, the Flathead National Forest's first telephone line linked the station with the forest supervisor's office forty-five miles away in Kalispell. But that original Spotted Bear Ranger Station didn't survive changes taking place on the forest.

A road completed along the east side of the South Fork of the Flathead River in 1924 resulted in construction of a second Spotted Bear Ranger Station on the south bank of the Spotted Bear River about a mile and a half east of the first. A new office building of lodgepole pine logs was built that year, and was followed in 1926 by a warehouse and in 1934 by a ranger's residence of nearly identical design and construction. These three buildings, finely crafted by Victor Holmlund, represent a rustic Rocky Mountain Cabin architectural style soon eclipsed by the style of the Civilian Conservation Corps "building boom" of the Great Depression. Although isolated examples of this log building style are found in the Region 1 backcountry, the Spotted Bear Ranger Station Historic District is the only example accessible by car and in continuous use since construction. These buildings, supplemented by tents on frames for seasonal employees, were typical of remote Northern Region ranger stations of the 1920s and 1930s.

The late Charlie Shaw, Spotted Bear Ranger Station fire dispatcher in 1929 and district ranger from 1945 to 1958, left vivid impressions of life there. The station was a center of Flathead National Forest horse and mule packing, the main way of supplying Forest Service backcountry operations before roads began to replace mountain trails. Packing wasn't always smooth going, and once in the late 1920s produced what Ranger Shaw called "the grandest mixup of all" at Spotted Bear Ranger Station. As he told it in *The Flathead Story:*

> Eleven full strings (10 head each) were packing out of there for the Spotted Bear, Black Bear, and Big Prairie districts. One morning, nine of these strings were going to leave the station for various destinations. The strings were packed and ready to leave at about the same time. Packstrings were tied up all over the place when Jack Langtree, the station cook, asked the packers to come in for a cup of coffee before they pulled out.
>
> While they were in the kitchen tent drinking their coffee, something "spooked" one string. They broke loose and started bucking, bawling, and running through the other eight strings. This caused every string to break apart and stampede. There were mules and horses bucking and running all over the Spotted Bear Ranger Station. . . . Ninety head of horses and mules were involved in this mixup. Some of them became tangled up. Some were down. But most of them stampeded into the woods, trailing packs and equipment as they ran.
>
> This happened about 9 a.m. But nobody got his outfit together that day. Mules and packs were found the next day as far away as Harrison Creek, eight miles south of the station. Some were found on Horse Ridge and in Twin Flats in the other direction. The next day, some were found still carrying their packs. Others lost their packs, saddles, and halters.

The packers had a big job putting that mixup right, but "the mess was eventually straightened out, and the packers were on their way. The packers, in a class by themselves, accepted the mishap as part of the work of moving Forest Service supplies over mountain trails."

Ranger Shaw was fire dispatcher at Spotted Bear Ranger Station when it was saved from the 101-day Sullivan Creek Fire that burned 35,000 acres of the Spotted Bear District in 1929. As he wrote in *The Flathead Story:*

> Nick Mamer piloted Regional Fire Chief Howard Flint over the Sullivan Creek Fire on the South Fork in 1929. When Howard saw that the Spotted Bear Ranger Station was in the path of the fire, he dropped a message to Charlie Hash, Flathead National Forest fire control officer: "Save Spotted Bear Ranger Station at any cost." Hash pulled 600 firefighters into the area, set up 20 Pacific Marine pumps, and the ranger station was saved.

That was in the early days of aircraft use in fire control, and helped prove their value. During the 1930s, the Civilian Conservation Corps built a landing field at Spotted Bear Ranger Station. At the height of the 1953 fire season, a Ford Trimotor suffered engine failure on takeoff there. The plane was demolished, but none of the crew or smokejumpers aboard was seriously injured.

Forest Service wilderness visionary Robert Marshall, among those responsible for the more than 100-million-acre National Wilderness Preservation System treasured by Americans today, occasionally stayed at Spotted Bear Ranger Station during his 1930s trips to adjacent national forest primitive areas later combined to form the million-acre Bob Marshall Wilderness named in his memory. The station has been the hub of wilderness management in the "Bob" since 1940.

Adapting historic Spotted Bear Ranger Station buildings to current needs will ensure their survival into the twenty-first century. Efforts to restore and interpret the station continue. In 1994, for example, Passport in Time volunteers helped Forest Service experts with structural replacement and interior renovation jobs, repaired windows, and painted. Volunteers also catalog and store historic artifacts and photographs for future use.

The name Spotted Bear, attached to a river, a lake, and a mountain as well as to the ranger station, is one of the oldest on the Flathead National Forest. Tradition has it that a guide and two miners, crossing the Flathead Range in 1861, saw a black bear with an unusual amount of white on its breast and underside while they were camped near what, from then on, has been called the Spotted Bear River.

ACCESS: Spotted Bear Ranger Station Historic District is 55 miles from Hungry Horse, Montana. Turn off U.S. Highway 2 at Hungry Horse and follow Forest Road 895, an improved light duty road that crosses Hungry Horse Dam and follows the western shore of Hungry Horse Reservoir and the South Fork of the Flathead River to Spotted Bear Ranger Station. Although there are no formal tours of the Historic District, visitors are welcome to look around on their own. On request, ranger station personnel will open the old office building for visitors.

LOCHSA RANGER STATION
Clearwater National Forest, Idaho
(1927)

A tale of two locations and three names complicates the early story of Lochsa Ranger Station. One of the West's most visitable historic ranger stations, it was dedicated on July 3 and opened to the public on July 4, 1976, as Lochsa Historical Ranger Station, a memorial to Forest Service history.

What in 1925 became Zion Creek Ranger Station, and in 1927 was renamed Lochsa Ranger Station, began in 1921 as Boulder Creek Ranger station on the old Selway National Forest. When construction of U.S. Highway 12 up the Lochsa River appeared imminent, and the Boulder Creek Ranger Station was in its path, Ranger Ralph Hand got orders from Forest Supervisor Frank Jefferson in Kooskia to move his ranger station about a mile down the Lochsa River to Zion Creek. A new four-room ranger station cabin, completed and occupied in 1925, was named Zion Creek Ranger Station, a name changed to Lochsa Ranger Station in 1927 because, as Ranger Hand explained, the Selway National Forest had too many ranger stations named for creeks. In 1926, the old Boulder Creek Ranger Station cabin was dismantled, its logs floated

The old Boulder Creek Ranger Station cabin, Lochsa Historical Ranger Station, Clearwater National Forest, Idaho.

a mile downriver to Zion Creek, and reassembled for use as a bunkhouse.

The highway that caused this move reached Lochsa Ranger Station, by then a Clearwater National Forest ranger district headquarters, thirty years later.

Except for the old Boulder Creek cabin, the log buildings of today's Lochsa Historical Ranger Station were built between 1928 and 1933 by the district rangers assigned at Lochsa during those years and their crews. The first begun and one of the last completed, the so-called combination building was constructed as four separate log cabins later joined under a common roof with a full-length covered porch across the front. The first of these cabins, a commissary, was begun in the spring of 1928 by Ranger Hand, Assistant Ranger George Case, and a smokechaser named Tom Mattison. "During the summer," according to former Lochsa District Ranger Louis F. Hartig's 1989 book, *Lochsa,* "Ranger Hand put up the first four or five rounds of logs himself after work hours, just to have something to do. . . ." The commissary was finished in

It took four days to skid this 375-pound ranger's dwelling bathtub twelve miles up the Lochsa River Trail to Lochsa Ranger Station. U.S. Forest Service photo.

December, and followed by an office unit built in 1929, a toolroom in 1930, and a kitchen in 1933 that completed the combination building after Ranger LeRoy Lewis had taken over the district. During these years, the alternate ranger dwelling—which served as both office and dwelling—was completed in 1931, and the 1925 cabin was dismantled in 1933 to make room for the two-story ranger dwelling occupied by Ranger and Mrs. Lewis the following summer. The building materials were cut locally or, along with many of the furnishings, packed up the twelve-mile Lochsa River Trail on the backs of horses and mules or skidded up the trail on stone boats. The district ranger's desk, for example, arrived on one side of a mule balanced by a large ham and two bales of hay on the other. And it took four days to skid the 375-pound bathtub for the ranger's dwelling up the trail.

The disastrous Pete King Fire that roared up Lochsa Canyon in August 1934 almost destroyed the newly-completed Lochsa Ranger Station. On the afternoon and evening of August 17, when flames

Lochsa Ranger Station after the 1934 Pete King Fire. Buildings (left to right) are the combination building, the wood shed (behind flag pole), the alternate ranger's dwelling (now the visitor information center), the ranger's dwelling, and the reassembled Boulder Creek cabin. U.S. Forest Service photo.

raged around and overhead and choking firefighters struggled to save the buildings and themselves, the station seemed doomed. But, once the flames had passed, according to firefighter Buck Weaver's recollection:

> Stumps and logs still smoldered, while every spear of grass and vegetation had turned to inky ashes. The only remaining piece of ground which the flames had not raked was the half acre on which the Ranger Station and other buildings stood, a silent monument to the courage and stoutheartedness of 115 men. . . .

The clearing of eight acres in 1931 to provide pasture for the station's horses and mules aided that valiant defense. But those horses and mules, "taken out of the barn while their corral and feed burned" and left with no grass to graze, had to be fed oatmeal from the kitchen until they were evacuated. When, at the height of the inferno, that kitchen caught fire, the flames were doused with coffee.

41

Lochsa Ranger Station survived the Pete King Fire—evidence of the month-long fire that ultimately blackened 241,000 acres is still visible in Lochsa Canyon—to serve as Lochsa District headquarters until 1957, a year after U.S. Highway 12 reached it. In 1957, Ranger Hartig—ranger of the Lochsa District from 1943 to 1966—and his wife moved to the new Kooskia Ranger Station, and the old Lochsa Ranger Station was used as a work center and guard station. As former Ranger Hartig wrote in *Lochsa:*

> With the Bureau of Public Roads survey crews and Forest Service people staying there during the field season, the station was quite active until the Lewis-Clark Highway was completed in 1962. The ranger dwelling was occupied year-round through 1963. When I left the district in 1966, four smokechasers were stationed there during the fire season and the district packer stayed there on rare occasions when he was not in the field.

Restoration of the buildings began in 1965 or 1966, and Lochsa Historical Ranger Station was dedicated in 1976.

Lochsa Historical Ranger Station is beautifully preserved as the back-country ranger station it was in the 1930s—and, indeed, remained until 1956. The ranger's dwelling, office, kitchen, and tool room are furnished and equipped as they were then, and the alternate ranger dwelling houses a visitor information center. And there's still no electricity! But the station is more than a museum. Lochsa District wilderness rangers and trail crews still live at the station during the summer, and the pack string used for wilderness trail maintenance is still kept in the corral—and adds to the visitor's feeling of stepping back in time.

ACCESS: Lochsa Historical Ranger Station is located on U.S. Highway 12, the scenic Lewis and Clark Highway, about 50 miles northeast of Kooskia, Idaho, and about 85 miles southwest of Lolo, Montana. It is open from 9 a.m. to 4 p.m. daily, Memorial Day through Labor Day. Both a free self-guided tour folder and helpful interpreters, many of them volunteer Forest Service retirees, are available in the visitor information center located in the alternate ranger dwelling just above the parking lot.

A mule string being packed in front of the combination building, Lochsa Ranger Station, during the old days (above, U.S. Forest Service photo), and the combination building today (below).

Ninemile Remount Depot and Ranger Station, Lolo National Forest, Montana. The visitor center is in the building at the right.

NINEMILE REMOUNT DEPOT & RANGER STATION
Lolo National Forest, Montana
(1934)

For almost a quarter century, from 1930 to 1953, Ninemile Remount Depot—established about five years before Ninemile Ranger Station was built at the same site—provided experienced packers and pack animals for fighting fires and supporting back-country work projects throughout the vast roadless reaches of the Northern Rockies. Today, at historic Ninemile Remount Depot and Ranger Station, the tradition of Forest Service packing lives on— and helps meet modern wilderness management challenges.

"When Congress set aside forest reserves in 1891," as the 1991 Forest Service publication *Miles By Mule*—quoted throughout— tells it so well, "people took horses and mules for granted much the way we take automobiles for granted today. The first forest rangers packed all of their belongings onto a horse or mule and

44

saddled up to travel to their assignments. . . . They gave little thought about the impossibility of managing the land without their animal partners."

But, while rangers and fire crews in the rugged and roadless national forests depended almost entirely on pack strings of horses and mules, times were changing elsewhere. "Trucks and tractors started to replace pack and draft animals on farms. And as automobiles and good roads increased, many people sold their saddle horses in favor of more modern transportation. Local supplies of pack and saddle stock started to disappear."

"Then in 1929, during an unusually bad fire season, the Forest Service awakened to a new challenge. With the mechanization of rural America, there were no longer enough pack horses and mules available for hire to supply the crews fighting forest fires in remote areas."

"Driven by the need to supply their own demand, the Forest Service established Ninemile Remount Depot to raise and train quality pack and saddle stock." From 1930 through 1953, a once run-down ranch on Ninemile Creek within the Lolo National Forest evolved into a unique Forest Service "ranch"—modeled after U.S. Army cavalry remount depots which supplied horses for troopers— that bred saddle horses for forest rangers and "maintained 20 pack trains of nine mules each," experienced packers, and transportation "ready to mobilize in minutes for firefighting" and to support backcountry work projects. "When not fighting fire, the mules packed supplies to trail crews and freighted construction materials to remote mountain tops to build fire lookouts. The depot's pack trains packed just about everything imaginable that is essential for managing a national forest."

"Professional packers at the Remount Depot developed packing equipment and perfected the art of packing to its highest level. They standardized methods and equipment so that any trained packer could pack all of the tools and supplies needed for a twenty-five person crew with just nine mules. Their methods were the most humane for keeping pack animals healthy."

Also during the 1930s, men at the 600-man Ninemile Civilian Conservation Corps Camp three miles north of the Remount Depot constructed many of the depot's roads, fences, and distinctive Cape

Ninemile Remount Depot and Ranger Station about 1935 (above) and the ranger station's Cape Cod style buildings under construction (below). U.S. Forest Service photos.

Cod style buildings as well as the adjacent Ninemile Ranger Station buildings of the same style that remain the headquarters of the Ninemile Ranger District.

Then, in the early 1940s, aerial firefighting was pioneered at Ninemile when an early smokejumper base was established at Camp Menard, about a mile north of the ranger station. Smokejumpers trained at Ninemile Valley's so-called airport pasture. Often, droning Ford Trimotor airplanes had to buzz the field to clear it of grazing mules before the smokejumpers could hit the silk. And there was another tie-in with the Remount Depot. About one mule per smokejumper was needed to pack out the smokejumper's gear, stashed along mountain trails after fires for pickup. Smokejumpers based at the Aerial Fire Depot in Missoula continue to practice at the airport pasture, where the Ninemile District still cuts hay and grazes livestock.

In 1953, the regional forester decided to close Ninemile Remount Depot because of the increasing effectiveness of aerial firefighting—smokejumpers and helicopters—and improved road access. Remount Depot facilities were incorporated into the Ninemile Ranger District operation the next year.

"Then in 1964, just as it started to look as though pack mules might become obsolete, Congress passed the Wilderness Act. This law directs the Forest Service to manage some special areas to preserve their wild character; no roads, no mechanized equipment are appropriate in wilderness." In these areas, horses and pack mules continue to play their historic roles.

Today, the Ninemile Remount Depot, a Ninemile Ranger District unit, is home to the Northern Region Pack Train of nine pack mules handled by two experienced Forest Service packers. In addition to providing packing services on national forests throughout the region, the pack train is used to help educate the public and Forest Service personnel on low-impact livestock use in wilderness and to represent the Forest Service at parades, fairs, rodeos, and other special events. A typical year might find the unit packing fifty thousand pounds of equipment and supplies—about 250 mule loads—over 750 trail miles into remote areas, participating in twenty parades and rodeos, and presenting twenty educational programs.

47

Ninemile Remount Depot mule strings were trucked to trailheads (above) where they hit the trails into the backcountry (below). U.S. Forest Service photos.

48

The art of Forest Service mule packing, as perfected at Ninemile Remount Depot (above, U.S. Forest Service photo), is exhibited at the visitor center (below).

At the Ninemile Remount Depot Visitor Center, which opened in 1989, visitors see how America's biggest pack station operated in the 1930s and 1940s.

Also at Ninemile Ranger Station are the Ninemile Wildlands Training Center and the Arthur H. Carhart National Wilderness Training Center for wilderness managers and wilderness rangers.

ACCESS: Ninemile Remount Depot and Ranger Station is less than an hour west of Missoula and just a short drive east from Alberton. From Missoula, take Interstate Highway 90 west for 23 miles to Exit 82; from Alberton, drive east for 12 miles to Exit 82. Follow Ninemile Road from Exit 82 for about 1 1/2 miles to Remount Road, then follow Remount Road about 2 1/2 miles to historic Ninemile Remount Depot and Ranger Station where the visitor center is open from mid-May through mid-September. A brochure available at the visitor center will guide you on a tour of the historic Remount Depot.

LUBY BAY RANGER RESIDENCE
Kaniksu National Forest, Idaho
(1935)

The old Luby Bay ranger residence—now the Priest Lake Museum and Information Center—affords an intimate glimpse of forest ranger life and work in the 1930s as it tells the wider tale of the Priest Lake region.

Built at Luby Bay on Priest Lake's western shore by the Civilian Conservation Corps in 1935, this cabin was one of several Kaniksu National Forest ranger stations and cabins—located a day's pack mule trip apart—that once ringed northern Idaho's famous nineteen-mile-long lake. And now, beautifully restored, it is the only one of those lakeshore facilities that remains.

Ranger James Ward—the first ranger on the Kaniksu National Forest, he served in the Forest Service from 1909 to 1939—and his family occupied the two-bedroom, one-bath log house when it was completed. Today, two rooms, the kitchen and the living room, are furnished as they might have been at the time. Among the furnish-

The old Luby Bay ranger's residence, Kaniksu National Forest, Idaho, is now the Priest Lake Museum and Information Center.

ings are a bearskin rug, donated by a Priest Lake woman who had three children and couldn't divide the rug among them, and a caribou head, shot by a member of President Theodore Roosevelt's hunting party when the hunters passed through the region in the early 1900s. Following its years as a district ranger's home, the cabin was used as a Forest Service guard station and again as a residence until 1989. In June, 1990, it opened as a museum and visitor center operated by the Priest Lake Museum Association and the Idaho Panhandle National Forests of which the Kaniksu National Forest is now a unit.

A large exhibit gallery in the cabin's bedrooms depicts many aspects of the Priest Lake region's natural and human history. Of particular interest is the major role boats played in the early years of the Forest Service in the Priest Lake country. Then, with few roads into the Kaniksu National Forest, water transportation was vital. At first, rangers patrolled in a rowboat fitted with a sail. Later, work boats served a variety of purposes from delivering mail and supplies to transporting fire crews, project crews, and even pack strings towed on barges. One of these work boats, confiscated by the U.S. Coast Guard from Puget Sound rum runners during the Prohibition era, was renamed *Kaniksu* and used by the Forest Service until 1948. Eventually, as roads afforded easier access to the forest, the lake was no longer key to transportation, the Forest Service no longer depended on work boats, and ranger stations were moved inland.

Also displayed in the Forest Service section of the museum is a Forest Service guard badge once worn by temporary Forest Service employees when only regular Forest Service officers wore the bronze pine tree shield. This rare nickel shield bears a pine tree flanked by the letters "U.S." and "F.S." and the words "Forest Guard" across the top and "Department of Agriculture" along the bottom.

ACCESS: Follow Idaho Highway 57 north from Priest River for 28 miles, turn right on Luby Bay Road, and proceed 1 1/2 miles to the junction of Luby Bay and Lakeshore roads. Turn left onto Lakeshore Road, proceed about 1/4 mile, and turn right at the Priest Lake Museum-Visitor Center sign. Volunteers staff the museum and visitor center from 10:00 a.m. to 4:30 p.m. daily from Memorial Day through Labor Day.

FENN RANGER STATION
Nez Perce National Forest, Idaho
(1939)

Half a million dollars was a big price to pay for a ranger station during the Great Depression—even one designed and built to accommodate two ranger districts. But that's what the government spent on Fenn Ranger Station, built on a bench above the Selway River by its Civilian Conservation Corps between 1936 and 1939. And, given that Fenn Ranger Station remains in use—as headquarters of the Selway Ranger District of the Nez Perce National Forest—after more than half a century, the government seems to

Fenn Ranger Station, Nez Perce National Forest, Idaho.

have gotten the taxpayers' money's worth. Today this beautiful cluster of buildings, surrounded by lawns and trees and serviced by sweeping curved drives, adds up to one of the Forest Service's classic historic ranger stations. It's a far cry from the first log cabin ranger station built just over the Bitterroot Range just forty years before.

Construction of Fenn Ranger Station began in 1936. The next year, an administration building, two warehouses, and two garages were completed. By the end of 1938, a cookhouse, gashouse, and one residence had been added. Another residence, a bunkhouse, and a powerhouse were completed in 1939. The barn was built in 1940.

By the fall of 1939, when Fenn Ranger Station was ready to occupy, Middle Fork and Selway ranger district personnel began moving in from the surrounding O'Hara, Pete King, and Number One ranger stations. Each district was given office space in the administration building, a warehouse, and a garage. Each district ranger had a house and a garage. The two districts shared meeting rooms, a cookhouse, a bunkhouse, and a gashouse. Electricity was generated by a Johnson Creek hydroelectric plant.

The move into Fenn Ranger Station was completed by the spring of 1940, just in time for one of the most severe fire seasons in Nez Perce and Clearwater national forest history. But that year those forests had more men to fight fires than ever before. Each of the Fenn Ranger Station districts staffed at least a dozen fire lookouts, and both had large Forest Service and CCC crews from which to draw firefighters. After the fall of 1940, when it appeared the United States eventually would enter the war in Europe, manpower at Fenn Ranger Station dwindled. Congress had passed the first peacetime draft, and the CCC camps and Forest Service crews began to shrink. Soon there was only one ranger at Fenn Ranger Station; he ran both districts during most of World War II. With a national forest reorganization in 1956, Fenn Ranger Station became a one-district station.

Fenn Ranger Station owes its name to local Forest Service rangers who admired and respected Major Frank A. Fenn, a distinguished Idaho pioneer and supervisor of both the Clearwater and Selway national forests. Chief Forester Gifford Pinchot called him "one of the very best forest officers the West has produced." The originally proposed name of Goddard Bar Ranger Station—that recognized the sloping, south-facing bench along the Selway River on which the station is sited—was opposed by local rangers on the grounds that Mr. Goddard had been less than an outstanding citizen. When the time came to order a sign for the new station, it was decided to change the name. In a letter to Regional Forester Evan W. Kelley, Nez Perce National Forest Supervisor Roy Phillips justified the change: "It is a well-known fact that no forest officer in these parts has ever been able to measure up to Major Fenn in the estimation of those who knew him."

Congress saved more than wild and scenic rivers when it passed the Wild and Scenic Rivers Act of 1968. It also saved Fenn Ranger Station. As part of a comprehensive Columbia Basin development plan, the U.S. Army Corps of Engineers in the early 1950s recommended building a dam on the Middle Fork River about four miles above Kooskia. Had the dam been built, Fenn Ranger Station would be under 100 to 300 feet of water and forty-eight miles of the then new Lewis and Clark Highway would have had to be relocated at a cost of almost forty million dollars. But the dam was never authorized. Wild rivers legislation was introduced in Congress in 1965, and supported by the Forest Service. Reintroduced and modified in the next Congress, it was passed and signed into law in October 1968. And historic Fenn Ranger Station remains one of the Forest Service's showplace ranger stations—a monument to Major Fenn and to the CCC men who built it.

ACCESS: Fenn Ranger Station is located on the beautiful Selway River about 5 miles southeast of Lowell, Idaho. Turn off U.S. Highway 12 just south of Lowell, cross the bridge over the Lochsa River just above its confluence with the Selway River, and follow paved Forest Road 223 about 5 miles to the ranger station on the left. The office is open daily, June through October. Fenn Pond, across Forest Road 223 from Fenn Ranger Station, has been developed with boardwalks and piers and is a nice fishing site for small children and others who have difficulty getting down to the Selway or Lochsa rivers. The site of O'Hara Ranger Station, 32 miles from the nearest road at Kooskia when it was built in 1910, is interpreted 2 1/4 miles up the road from Fenn Ranger Station.

DARBY RANGER STATION
Bitterroot National Forest, Montana
(1939)

The old Darby Ranger Station building—called Darby Historic Ranger Station since the National Forest System Centennial in 1991—was the first ranger station in Darby. Before it was built, Forest Service activities were conducted at rangers' homes and in an office above the bar in the Darmont Hotel.

Darby Ranger Station, located on a ten-acre site purchased in 1936 for five hundred dollars, was built between 1937 and 1939 by the Civilian Conservation Corps and local craftsmen. From its completion until 1964, the office building that now houses an interpretive and information center was headquarters of the

Darby Historic Ranger Station,
Bitterroot National Forest, Montana.

358,000-acre Darby Ranger District of the Bitterroot National Forest. The district ranger's offices are now in the newer building next door. From 1975 until 1990, the historic ranger station office building was headquarters of the Bitterroot Hotshots, an elite Forest Service interregional fire crew. Since 1991, it has been developed as a visitor and historic center staffed by dedicated volunteers. Among excellent exhibits that depict local Forest Service history are one about Ranger Than Wilkerson of Alta Ranger Station fame, a 1930s-era ranger's office, and a tool and tack room.

ACCESS: *Darby Historic Ranger Station is conveniently located at 712 U.S. Highway 93 North in Darby, Montana, and is open from 9 a.m. to 5 p.m. seven days a week from Memorial Day through the fall hunting season.*

Chapter Two
ROCKY MOUNTAIN REGION

The seventeen national forests of the Rocky Mountain Region, or Region 2, encompass about twenty million acres of the Rocky Mountain ranges in Wyoming and Colorado, the Black Hills of South Dakota, and the Sandhills of central Nebraska. Astride the Continental Divide, the fifteen Rocky Mountain national forests are the heart of this region. Within their boundaries are more than a thousand peaks and plateaus over ten thousand feet high, more than fifty that rise more than fourteen thousand feet. Their forested slopes are clothed mainly by pine, fir, and spruce. East of the Rockies are the pine-clad Black Hills National Forest and the Nebraska National Forest, the nation's largest hand-planted forest. These forests are managed by a regional forester near Denver, eleven forest supervisors, and about sixty district rangers.

Region 2 is the birthplace of the National Forest System and the first forest ranger. President Benjamin Harrison proclaimed the Yellowstone Timberland Reserve—of which the region's Shoshone National Forest is a descendant—on March 30, 1891. Just over seven years later, on August 8, 1898, young William R. "Bill" Kreutzer of Colorado was appointed the first U.S. forest ranger. He made the transition to the Forest Service in 1905 and served as a ranger, forest supervisor, and district forest inspector before he retired in 1939. The region's first ranger station, Wapiti Ranger Station built in 1903 on what soon became the Shoshone National Forest, remains in service, and is one of thirteen historic ranger and guard stations profiled in this chapter.

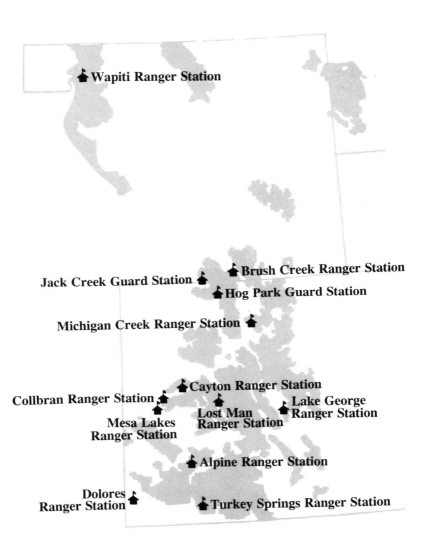

Thirteen historic U.S. Forest Service ranger stations of the
Rocky Mountain Region (Region 2) are in Colorado and Wyoming.

Historic Wapiti Ranger Station, Shoshone National Forest, Wyoming, in May 1968.

WAPITI RANGER STATION
Shoshone National Forest, Wyoming
(1903)

America's first national forest claims one of America's first ranger stations. Shoshone National Forest, carved out of lands set aside by President Benjamin Harrison as the Yellowstone Timberland Reserve in 1891 to be the first unit of the National Forest System, is the site of Wapiti Ranger Station, *the first ranger station built with government funds and still in service at its original location.*

Other ranger stations were built before Rangers Harry Thurston and Milton Benedict erected Wapiti in 1903. Alta Ranger Station was built in 1899 on the Bitter Root Forest Reserve, but with the rangers' personal funds (pages 14 to 18). And West Fork Ranger Station, built in 1900 on the San Gabriel Forest Reserve, was the first ranger station in California built with federal funds, but has been moved almost ten miles from its original location and reconstructed at a visitor center (pages 152 to 155).

At any rate, it was October 1903 when Ranger Thurston rendezvoused with Ranger Benedict and his sixteen packhorse loads of supplies at the confluence of the North Fork and Elk Fork of the Shoshone River to construct quarters for W.H. Pierce, supervisor of the Shoshone Division of the Reserve, who had decided to relocate his headquarters from Cody. "The site," retired Forest Supervisor Thurston wrote in 1941, "was delightfully located in a broad stretch of mountain valley. Water was near, wood abundant, and fish and game plentiful." The two rangers built a three-room dwelling and a separate office cabin of logs previously "cut, peeled, and adzed by Ranger Sherwood." The space between the two buildings was later enclosed. In 1904, the first telephone line built by the government for administration and protection of forest reserves linked Wapiti with Cody, and the stable and corrals were built. Supervisor Pierce and Ranger Thurston lived and worked there until 1907. Distinguished visitors to Wapiti during those first years included Colonel William F. "Buffalo Bill" Cody and Chief Forester Gifford Pinchot. When, in 1907, the divisions of the Reserve were separated and designated "national forests," Shoshone National Forest headquarters moved to Cody and Wapiti Ranger Station became headquarters of its North Fork Ranger District.

After the North Fork and Canyon Creek districts were combined to form the Wapiti Ranger District in 1921, Ranger Clifford Spencer and his wife, Helen, moved to Wapiti Ranger Station from flood-ravaged Canyon Creek Ranger Station. Wapiti had been unoccupied for some time when the Spencers moved in during the summer of 1923. After cleaning, repairs including placing a new foundation under the building, and adding a bedroom, the Spencers were at home.

Helen Spencer, "practically an assistant ranger" to her husband, recalled her life as a Forest Service wife at Wapiti Ranger Station in a 1979 interview. In addition to chopping wood, carrying water, providing information to visitors, lining out fire crews, and myriad other tasks, Mrs. Spencer's specialty seemed to be catching and saddling the horses for Ranger Spencer. "If he wanted the horses in a hurry," she explained, "I (would) rattle the oat can. They would come tearing to the house. When he did it, the horses decided he wanted to use them, so they wouldn't come." Ranger Spencer did a lot of his work from horseback, and Mrs. Spencer accompanied him on many of his pack trips.

> I was his trail crew. The Forest Service didn't get any money in those days to hire a trail crew, so I went along to help. He always took along a two-handled saw so I could get on the other end of it.

Often, when the Spencers crossed the road between Cody and Yellowstone National Park with a pack string, tourists stopped to take their pictures.

And, as was Forest Service custom, Mrs. Spencer fed all comers—from Washington, D.C., big shots to a Civilian Conservation Corps crew on a bug job to Great Depression victims who would "stop in and want a meal." Among the Spencers' better known Wapiti guests were Yellowstone National Park Superintendent Horace W. Albright, who became National Park Service director, and Department of Agriculture entomologist Dr. F.C. Craighead and his twin sons, Frank and John, who became famous grizzly bear researchers. Otherwise, social life was limited to dances at Wapiti Schoolhouse at which ranger station and neighboring ranch couples celebrated birthdays and wedding anniversaries.

Although she summed up her Wapiti Ranger Station years as "an enjoyable time," Mrs. Spencer recalled some tense moments. Once, on a day after Christmas when she was alone at the station, she answered the knock of a drunken man who claimed "he had just shot a man at the sawmill" and was on the run from the sheriff. "You're in a helluva fix, aren't ya?" he menaced. "I could hurt you if I wanted to . . . but I won't." He borrowed a can opener, ate his canned sardine lunch in the garage, returned the can opener, and helped himself to oats for his horses before he left. "He was so drunk," Mrs. Spencer recalled, "that when Clifford came home he wanted to know what kind of party I'd been having in the garage. It smelled so terribly of moonshine." The sheriff caught up with and arrested the fugitive, who had shot two fingers off his victim's hand during a Christmas Day quarrel, the next day.

The original Wapiti Ranger Station building was improved over the years. When, toward the end of the Spencer's years at Wapiti, the CCC added a long-hoped-for fireplace and a bathroom, Mrs. Spencer "took a bath about three times a day . . . it was such a treat to have a bath tub and play in hot water." After Ranger Spencer was transferred to Colorado in 1937, two bedrooms were added in the early 1940s and a utility room was added in the early 1950s.

Wapiti Ranger Station remained a ranger district headquarters until 1957, when the Wapiti and South Fork districts were combined into a larger Wapiti Ranger District and the office moved to Cody. The historic ranger station continues in use as a guard station and work center from which campgrounds are maintained and at which pack animals—horses and mules—are kept.

ACCESS: Historic Wapiti Ranger Station is located on the north side of U.S. Highway 14-16-20 about 25 miles west of Cody and 30 miles east of the Yellowstone National Park boundary. Although still used as a residence and not open for public visitation, the station is easily viewed from a visitor center adjacent to the highway and staffed during the summer.

ALPINE RANGER STATION
Uncompahgre National Forest, Colorado
(1907)

Still in use as Alpine Guard Station, historic Alpine Ranger Station consists of three buildings—a residence, a garage, and a barn—built between 1907 and the 1930s.

The original one-room log cabin dwelling, built by forest rangers in 1907, was declared "entirely inadequate" by District Forest Inspector William R. Kreutzer a dozen or so years later, and by 1920 construction of a new residence was underway. At some point, the 1907 cabin was converted to the log garage it is today by installation of a garage door.

Alpine Ranger Station, Uncompahgre National Forest, Colorado, in the early 1920s. U.S. Forest Service photo.

Ranger William Doran, in a November 4, 1920, report to the Uncompahgre National Forest supervisor, reported completion of the new residence at Alpine Ranger Station. This building, of Engelmann spruce logs and materials salvaged from the demolition of several buildings in nearby Lake City, was built by contractor George Vernon under Ranger Doran's supervision for $894.19. The new dwelling had log walls, a gabled roof, and a wide porch that ran the full length of the northwest side. The porch was later replaced by a small stoop and shed-roofed porch. A new concrete foundation and new sill logs were installed in 1984.

The barn was built during the Great Depression by the Civilian Conservation Corps. Despite several attempts over the years to replace them, all three historic Alpine Ranger Station buildings remain in regular service.

ACCESS: It's about 23 1/2 dirt road miles from U.S. Highway 50 to Alpine Guard Station. To get there, turn south off U.S. Highway 50 about 10 miles east of Cimarron or about 30 miles west of

Gunnison onto BLM Route 3004, which follows the East Fork of Little Blue Creek southward for more than half the 13 1/2 miles to the Uncompahgre National Forest boundary where BLM Route 3004 becomes Forest Road 867. Follow Forest Road 867 for about 5 1/2 miles to its junction with Forest Road 868, the Alpine Road. Turn right onto and follow Forest Road 868 about 4 1/2 miles to Alpine Guard Station, just north of Big Blue Campground.

CAYTON RANGER STATION
White River National Forest, Colorado
(1910)

It was called Johnson Spring Ranger Station in 1909 and 1910 when Ranger James G. "Jim" Cayton and his new bride, Adelaide D. "Birdie" Miller of Silt, Colorado, and other Forest Service rangers, began to build the three-room log house and log barn from which he ran his district until 1919. Ranger Cayton's own words, written thirty years later, tell it best.

"In September 1909 Forest Ranger Jolly Boone Robinson and I first started the improvements at this station. They consisted of a log barn and a three-room log house. After starting Ranger Robinson at cutting and peeling the green Colorado blue spruce trees for the logs along West Divide Creek, I went out on a trip doing my ranger district work for a few days.

"When I returned I took my bride of just a few days to the station site, where we lived in tents, cooking over a camp fire, then later on an old cook stove. One evening Mrs. Cayton made the remark that if she had a rolling pin, she would build us a pie. Boone picked up a small green aspen stick and proceeded to whittle out a rolling pin with his pocket knife. We had the pie the next day. . . .

"We built the barn first and put the shingle roof on it, then moved into it as there was nearly two feet of snow on the ground and snowing most of the time. We chinked the barn, then dug a hole in the dirt floor, mixed the mud and daubed it on the inside. The barn made quite comfortable living quarters as compared to the tents.

Historic Cayton Ranger Station, Grand Mesa National Forest, Colorado, in 1993. National Park Service photo by Midwest Archaeological Center.

"We laid up four rounds of the logs for the house. Then discontinued work on it for the winter. Next spring Rangers Robinson and Jack Hughes came and we completed laying up the logs for the house, with log partitions, got the shingle roof on it, and put in the doors and windows, when Robinson and Hughes had to return to their ranger districts.

"During that summer season of 1910 Mrs. Cayton and I put the chinking in the house and daubed it with mud. We also built the brick chimney, she being the hod carrier. The hole had been cut in the roof for the chimney and in order to make the chimney go out the hole (this was the first brick chimney I had ever built or seen built) I had to make about two curves in it. The first time Supervisor John W. Lowell, Jr., came after the chimney was built, he looked at it and said to me 'Jim, I understand now why the smoke curls so nicely when it comes out of your chimney, it's because of the artistic curves you put in it.'

Cayton Ranger Station, then called Johnson Spring Ranger Station, in 1916 when Ranger and Mrs. James G. Cayton lived there (above), and Cayton Ranger Station in 1939 (below). Photos from the album of Mrs. James G. Cayton, courtesy of White River National Forest.

"During the summer of 1910 Mrs. Cayton and I moved into the Ranger Station, making it a year around headquarters from then until 1919 when I resigned and we went to California for her health."

During those years, the Caytons made Johnson Spring Ranger Station their "happy home." They fenced a half-acre yard around the house and seeded it to Kentucky blue grass for a lawn. They transplanted small aspens to the yard. And, in the spring of 1919, they buried Curly, their little dog who kept Mrs. Cayton company when Ranger Cayton was out on his district, inside the yard where they also "planted memorial trees for three of the Forest Service boys who were killed in action in France" during World War I.

Ranger Cayton left the Forest Service and Johnson Spring Ranger Station in 1919 because of Mrs. Cayton's health. He took her to Gilroy, California—just south of San Jose, bought a candy and ice cream store, and became a master candy maker. Mrs. Cayton recovered, and in 1920 the Caytons returned to the Forest Service and Colorado. Ranger Cayton was assigned to Rico Ranger Station on the old Montezuma National Forest, now part of the San Juan National Forest, for ten years. In 1930, he was moved to Rifle, Colorado, where he served until his 1939 retirement. That year, in his honor, Regional Forester Colonel A.S. Peck renamed Johnson Spring Ranger Station, where the Caytons had lived and worked from 1910 to 1919, Cayton Ranger Station.

Born in Hooper, Nebraska, in 1878, Ranger Cayton had worked at ranching, mining, and on the Gunnison Tunnel before being appointed a Department of the Interior forester in 1903, and a forest ranger in 1904. He joined Gifford Pinchot's new Forest Service in 1905. A dedicated public servant, he lived the rugged life of the early forest ranger. Following his retirement, the Caytons lived in Rifle. During World War II, retired Ranger Cayton served in the U.S. Army's auxiliary police and as a guard at strategic Colorado defense plants. He later became a police magistrate, notary public, and justice of the peace. James G. Cayton, pioneer forest ranger, died in 1956.

ACCESS: Historic Cayton Ranger Station, off West Divide Creek Road about 4 miles south of the Garfield County-Mesa County line, is about 26 road miles southeast of Rifle. To get there, exit Interstate

Ranger James G. Cayton on West Divide Creek, White River National Forest, Colorado. Photo from the album of Mrs. James G. Cayton, courtesy of White River National Forest.

Highway 70 at Rifle, turning south onto Garfield County Road 346, and drive east on that road for about 7 miles to Garfield County Road 31. Or, exit Interstate Highway 70 at Silt, cross the Colorado River to Garfield County Road 346, and drive west about 1 mile to Garfield County Road 331. Follow Garfield County Road 331, which ascends Dry Hollow Creek, south about 9 miles, then east about 1 1/4 miles to Garfield County Road 342. Follow this road south for 1 mile, then turn east on Garfield County Road 344. After about 1 mile, this road turns southwesterly to ascend West Divide Creek and cross into Mesa County. After about 7 1/2 miles, Garfield County Road 344 crosses the White River National Forest boundary and becomes Forest Road 800. Cayton Guard Station is west of this road about 1 mile past the forest boundary. The last 5 or 6 miles of this trip is on unimproved roads that may be impassable when wet, so this trip should be attempted only during dry weather and road conditions by travelers who have a White River National Forest map.

Hog Park Guard Station, Routt National Forest, Colorado, in 1993.
National Park Service photo by Midwest Archaeological Center.

HOG PARK GUARD STATION
Routt National Forest, Colorado
(1910)

Hog Park Guard Station, a log cabin and barn built by the Forest Service between 1910 and 1912, is the oldest remaining guard station on the Routt National Forest. Less than one-half mile south of the Colorado-Wyoming state line and a couple miles east of the Continental Divide, this remote historic guard station is sited on a narrow, forested terrace on the western bank of the Encampment River.

Forest Supervisor Jesse W. Nelson of the old Sierra Madre National Forest requested $450 in 1907 to build a cabin, barn, and fencing at Hog Park as a base for rangers responsible for overseeing settlement of a large timber trespass by the Carbon Timber Company as well as managing forest lands and enforcing federal regulations. By the time the cabin and barn were completed in 1912, the forest's name had been changed to Hayden National Forest. There was another name change in 1929, when the Colorado part of the Hayden National Forest became part of the Routt National Forest and the Wyoming part was absorbed by the Medicine Bow National Forest.

Ranger Richard Riggs and several other Forest Service employees spent the winters of 1910-1911 and 1911-1912 at the station, which must have been mostly completed by 1910 even if not reported fully completed until 1912. Records show that Ranger Reed Higby and his wife served at the station during the summers of 1918 through 1923.

Cutting railroad ties, an activity that peaked in the Hog Park area between 1900 and 1910, seems to have stimulated establishment of Hog Park Guard Station. According to the September 18, 1902, issue of *The Denver Times*, the Carbon Timber Company—the company later charged with trespass—had some three hundred "tie hacks"—men who cut and shaped railroad ties with hand tools—and other workers in the area to supply ties, mine timbers, and poles to the Union Pacific Railroad in Wyoming. But, by the time the guard station was completed, the timber company had ended its tie cutting operation.

Still in service, historic Hog Park Guard Station supports Routt National Forest operations. Both its 15-foot by 29-foot cabin and 13-foot by 17-foot barn are built of round logs with square-notched, smooth-sawn ends, and have wood-frame gables and roofs finished with wood shingles. Surrounded by a stand of lodgepole pine, Englemann spruce, and Douglas-fir, their appearance has changed little since they were built years before the Forest Service developed standard building plans.

ACCESS: Remote is the word for historic Hog Park Guard Station. To get there, turn off Colorado Highway 125 at Cowdrey, about 9 miles north of Walden and about 58 miles south of Saratoga, Wyoming, onto Jackson County Road 6W, a paved road that becomes an all-weather road after about 5 miles. Follow Jackson County Road 6W for about 16 miles to the Routt National Forest boundary, where it becomes Forest Road 80. Follow all-weather Forest Road 80 for about 18 more miles to its junction with the historic Ellis Trail, which provides vehicle access to Hog Park Guard Station. Turn south on the Ellis Trail to the historic station on the western bank of the Encampment River 1/4 mile south of Forest Road 80.

Lost Man Ranger Station, White River National Forest, in 1940.
U.S. Forest Service photo.

LOST MAN RANGER STATION
White River National Forest, Colorado
(1913)

The small log cabin once called Lost Man Ranger Station reflects the Forest Service's early appreciation of high-speed communications in national forest administration. The station was built in October 1913 by the Forest Service in cooperation with the Mountain States Telephone Company primarily as a line cabin for crews who maintained telephone lines over the Continental Divide just a few miles to the east. Records show the Forest Service provided $85.04 in labor to the project while the telephone company supplied $50.00 worth of materials.

Telephones were in use at ranger stations from the earliest days of the Forest Service, which both built its own telephone lines and contracted with private telephone companies—often trading resources for access. Rangers frequently took on the responsibility of maintaining the lines they used. This assured not only their continued use of the line, but the good will of the telephone company and other users.

By the 1940s, Lost Man Ranger Station, later called Lost Man Guard Station, was used only irregularly by Forest Service personnel as a temporary stopping place. Today, the historic Lost Man Ranger Station cabin reminds visitors of earlier eras in both communications technology and national forest administration.

ACCESS: The historic Lost Man Ranger Station cabin is on the south side of Colorado Highway 82, adjacent to Lost Man Campground, about 12 miles southeast of Aspen and about 6 miles west of Independence Pass.

TURKEY SPRINGS RANGER STATION
San Juan National Forest, Colorado
(1929)

Sixty-five years after Ranger John Baird and his wife, Sally, were wed in Pagosa Springs, Colorado, on June 18, 1929, they recalled their honeymoon at Turkey Springs Ranger Station—and how the ranger station burned to the ground as they cooked the first breakfast of their new life together.

The stove, they explained, caught the roof afire. The fire was soon out of control. Nearby sheepherders, who saw the flames and smoke, came to lend a hand. They helped rescue all the wedding presents, then the other objects from the cabin. Last out was the telephone. Ranger Baird removed it after calling neighboring Ranger Ray Taylor to explain that the ranger station was burning down around him. Not wanting to miss a free meal, the sheepherders even rescued the stove, with fire and breakfast still in it, and carried it to the lawn. All present had a picnic breakfast while the cabin burned.

The cabin that burned that first day of summer in 1929 had been built in 1921. It was one of four on the San Juan National Forest ranger district to which Ranger Baird had been assigned following his first two Forest Service years on the Medicine Bow and

Historic Turkey Springs Ranger Station, San Juan National Forest, Colorado. National Park Service photo by Midwest Archaeological Center.

Shoshone national forests in Wyoming. He'd gone to college in Idaho, studied forestry and range management, and accepted the transfer because it offered more of the range work he wanted. He'd come to the right place. His predecessor had set up housekeeping at Turkey Springs Ranger Station to count sheep onto the range and do other jobs to improve the overgrazed rangelands. The trail to the high country ranges ran through Turkey Springs, and there were about twenty sheep camps on the district. Rangers and their families lived in these stations close to their jobs.

Ranger Baird rebuilt the Turkey Springs Ranger Station cabin in 1930. After his sketched plans were carefully drawn up by the district—now regional—office engineers in Denver, he confessed in a 1994 interview, he ignored the plans and just built the cabin as any building should be built. It was later redesignated a guard station, and later still used by range riders working for ranchers who grazed cattle on national forest rangelands. Today, Ranger Baird's historic Turkey Springs cabin is a San Juan National Forest recreation rental cabin at which visitors may experience something of early ranger life.

ACCESS: To reach historic Turkey Springs Ranger Station, turn north off U.S. Highway 160 onto Forest Road 631, Piedra Road, about 2 miles west of Pagosa Springs. Drive north on Piedra Road for 6 1/2 miles to its junction with Forest Road 629, Turkey Springs Road. Turn left onto Turkey Springs Road and travel 3 miles to the right turn to the cabin indicated by a sign. After the turn, the road divides; stay to the left and reach the cabin 1/2 mile from Turkey Springs Road. Turkey Springs Road is closed seasonally to protect it from damage during the winter and spring, so check at Pagosa Springs Ranger Station for off-season road conditions. When the gate is closed, access is by cross-country skis, snowshoes, or snowmobiles (with adequate snow). In spring and late fall, access may be by foot only. Parking is available on Piedra Road just below its junction with Turkey Springs Road. Contact Pagosa Ranger District, P.O. Box 310, Pagosa Springs, Colorado 81147, telephone (303) 264-2268 for rental information.

JACK CREEK GUARD STATION
Medicine Bow National Forest, Wyoming
(1934)

In 1933, Ranger Evan J. "Evy" Williams, district ranger of the Medicine Bow National Forest's old Encampment Ranger District, began to build a log cabin guard station. With an allocation of $500, he purchased and delivered cement and other building materials, and hired a crew to fell and haul logs. This initial work exceeded Ranger Williams's $500 allocation by $75, and he had to complete the cabin on his own. With some help from another ranger, he finished it in 1934.

Jack Creek Guard Station has been used for fire, timber, range, and wildlife management work ever since. In six decades it has seen and been a part of the Forest Service's transition from foot and horseback trail patrols to automobile travel on well engineered

Jack Creek Guard Station, Medicine Bow National Forest, Wyoming, about 1935. U.S. Forest Service photo.

An interpretive sign tells the story of historic Jack Creek Guard Station, Medicine Bow National Forest, Wyoming. U.S. Forest Service photo.

roads. Down through those years, Ranger Williams and countless other Forest Service employees have relaxed on its covered porch after a hard day's work and watched the sun turn the country gold as it set behind the Continental Divide before turning in for the night. A single room, 14-foot by 16-foot, standard Rocky Mountain style cabin of peeled logs, tongue-and-groove floor, and shake roof with a yellow brick stone chimney and an outhouse out back, Jack Creek Guard Station wasn't a luxurious accommodation. But it did the job and, along with a new work center nearby, it still does.

Ranger Williams, who served his entire 35-year Forest Service career on the Medicine Bow National Forest, worked from the early years into modern times. He began in 1916, and retired in 1950. During that time, in addition to doing a ranger's normal duties, he built not only Jack Creek Guard Station, but also built or helped build or rebuild five ranger stations, two fire lookouts, a hundred miles of trail, fifty miles of telephone line, and fifty miles of boundary fence. But historic Jack Creek Guard Station is his monument—and his final resting place. About twenty yards north of the Jack Creek cabin is Ranger Evan John Williams's grave, marked by a triangular log fence and a large, natural, unmarked stone. The grave contains his ashes, buried there in 1970.

ACCESS: To reach historic Jack Creek Guard Station, follow Carbon County Road 500 west from Saratoga for 16 miles to the Medicine Bow National Forest boundary and Forest Road 452. Drive south on Forest Road 452 for 5 miles, past the Jack Creek Campground, to Jack Creek Guard Station which is about 15 yards north of the road.

MICHIGAN CREEK RANGER STATION
Routt National Forest, Colorado
(1937)

Since 1914, when a two-room log cabin and a log barn were built about eighteen miles southeast of Walden, Colorado, to oversee part of the North Park Ranger District, there's been a Forest Service station along the upper reaches of the Michigan River. Over the years, this station has been known variously as Michigan Ranger Station, South Michigan Ranger Station, and—inexplicably—Michigan Creek Ranger Station.

The carefully-crafted 1914 log cabin was improved by Civilian Conservation Corps workers in 1933. Then, in 1936, Ranger Carl Sward, who lived at what was then called Michigan Creek Ranger Station to administer a large timber sale, decided to get married. To enlarge his quarters, the Rocky Mountain Region's engineers in Denver drew up plans for a two-room addition to the cabin's west side. The addition was built in 1937. Another addition, planned in 1939, was not built. The Forest Service planned to use the cabin as a full-time ranger's residence during the timber sale, and then as a guard station.

That plan worked. The cabin remains in good condition and in seasonal use as Michigan River Guard Station. The original shake roof has been covered by a metal roof. The barn no longer exists.

Michigan Creek Ranger Station, Routt National Forest, Colorado, in 1938. U.S. Forest Service photo.

ACCESS: *The historic Michigan Creek Ranger Station cabin is about 1 mile off Colorado Highway 14, 23 road miles southeast of Walden and 77 road miles west of Fort Collins. Turn off Colorado Highway 14 near Gould onto Forest Road 740, and drive toward Aspen Campground. After about a mile, turn right off Forest Road 740 and follow the road across the South Fork of the Michigan River for about 1/2 mile to Michigan River Guard Station.*

COLLBRAN RANGER STATION
Grand Mesa National Forest, Colorado
(1937)

Back in 1937, long before the streets in the western Colorado cow-town of Collbran were paved, the Forest Service built Collbran Ranger Station there. Supervised by Ranger Arthur J. Ebert, Depression-era laborers "borrowed" by Grand Mesa National

Collbran Ranger Station, Grand Mesa National Forest, in 1940. Buildings (left to right) are the garage-shop, ranger's residence, and office (above, U.S. Forest Service photo). Collbran Ranger Station office today (below).

Forest Supervisor Ray Peck from construction of the beautiful Lands End Shelter House—now the Lands End Visitor Center—at the end of the Grand Mesa's challenging Lands End Road, constructed three wood-frame buildings—an office, a residence, and a garage-shop—on natural stone foundations. These buildings, headquarters of the Collbran Ranger District for almost six decades, were remodeled in the 1980s in a way that preserved their original appearance.

Although, as a result of a recent ranger district consolidation, the district ranger is now based in Grand Junction, historic Collbran Ranger Station remains in service.

ACCESS: Collbran Ranger Station, at 212 East High Street in Collbran, is open during normal business hours. To get there from the west, turn off Interstate Highway 70 onto Colorado Highway 65 about 20 miles east of Grand Junction, and follow that highway for about 10 miles to its junction with Colorado Highway 330. Turn onto Colorado Highway 330, and follow it for another 11 miles to Collbran. From the east, turn off Interstate Highway 70 at DeBeque and follow the paved DeBeque Cutoff Road for about 12 miles to its junction with Colorado Highway 65, turn left and cross Plateau Creek on that highway, then turn left again onto Colorado Highway 330 and drive 11 miles to Collbran.

DOLORES RANGER STATION
San Juan National Forest, Colorado
(1938)

The four cream-colored, adobe-style buildings of Dolores Ranger Station, built by the Civilian Conservation Corps between 1937 and 1940 as headquarters for two ranger districts on the old Montezuma National Forest, blend well with the surrounding sandstone cliffs. Now well into their fifth decade, these buildings on the corner of 6th Street and Central Avenue in Dolores, Colorado, still house the offices of the Dolores Ranger District on the San Juan National Forest.

Dolores Ranger Station, San Juan National Forest, Colorado, in 1947 (above, U.S. Forest Service photo) and about 1990 (photo by Milt Griffith).

81

Ranger Cliff Chappell of the old Glade Ranger District had been in Dolores since 1935. He and his wife, Ruth, had lived in a rented house in what Mrs. Chappell later called the "small, rough cowboy town" and watched construction of the new ranger station. Mrs. Chappell termed their February 1939 move into the first of the four buildings—the large corner building that now houses the district ranger's office—the "thrill of a lifetime" in a 1992 interview. "There were three bedrooms and . . . just the two of us. We thought we would close off some of the bedrooms," she recalled, but the forest supervisor was so proud of the new building he furnished every room. Later, the Dolores District ranger and family moved in next door. The two district rangers shared the office and warehouse.

But the Chappells weren't in town—and in their Dolores Ranger Station quarters—more than five months a year. Mrs. Chappell explained why:

> The rules were very strict. A ranger couldn't be off his district. Dolores was off of Cliff's district, so he couldn't come to town. He would hardly be able to get his haircut in the summertime.

Ranger Chappell had to be on his district during the field season to manage its timber and range resources and control forest fires. So the Chappells spent at least half the year at small "summer stations" like Glade Ranger Station. Mrs. Chappell had to make the trips into Dolores to buy supplies.

> When I came to town for groceries, I would come down in early morning and wait until after dark to avoid logging trucks. I would still meet logging trucks. I could see the lights far enough ahead that I could get to a turnoff area.

Back in Dolores during the winters, Mrs. Chappell helped Ranger Chappell with his district paperwork. "I remember that I spent days and days typing his forms. Neither of us were typists, but we would get it done." Ranger Chappell had no district clerk in those days. His staff was one summer assistant who helped with the field work.

Today, the office and two residence buildings at Dolores Ranger Station provide offices for the district ranger and his staff. While resource specialists at computers have replaced lone rangers at manual typewriters inside, about the only change in the historic ranger station's outward appearance is the pitched metal roofs added over the original flat roofs in 1984.

ACCESS: Dolores Ranger Station, on the corner of 6th Street and Central Avenue in Dolores, is about 10 miles north of Cortez on Colorado Highway 145.

LAKE GEORGE RANGER STATION
Pike National Forest, Colorado
(1939)

Three log buildings in the mountains west of Colorado Springs—the office, residence, and garage buildings of rustic Lake George Ranger Station, now a Forest Service work center—provide a beautiful example of the Rocky Mountain Region's Depression-era administrative architecture and Civilian Conservation Corps craftsmanship. These buildings replaced the previous Lake George Ranger Station building that was sold and reportedly moved northwest up Tarryall Creek to the Ute Creek Trading Post where it is said to be the largest of the cabins. As interesting as these log buildings is the story of their construction.

"All of the new buildings at Lake George were started after my transfer to the district . . . late in 1938," recalled former Ranger Wendell R. Becton of Gainesville, Georgia, in a 1990 letter.

Construction began during the early part of 1939. . . . There might have been some excavation for the basement during the latter part of 1938. As I remember, the residence was the first [building] to be completed . . . followed by the . . . office and barn. All labor was by the CCC boys. It required an average of one man day for each log. All of them had to be virtually hand carved to fit into its place and be tight against the next one below and above. A V-shaped trench was dug on the underside for the entire length of each log into which the oakum was placed to provide for as close an air tight as possible fit. At the corners,

83

Office building, Lake George Ranger Station, Pike National Forest, Colorado, in 1941. U.S. Forest Service photo.

each log was cut out to fit the next one below. It is easy to understand why so much hand labor was required. The foreman was unusually exacting with all the fits.

The logs for the buildings were provided by Charlie Heitz, who salvaged timber from the Hourglass Burn—a large forest fire on the Pike National Forest in the 1930s—and were hauled to the Lake George site by Bill Smith. At the site, a Forest Service construction foreman—remembered by Bill O'Shia of Portland, Oregon, who worked on the CCC crew that built the Lake George Ranger Station buildings, as "Mr. Blevins" and by former Ranger Becton as "an expert in log house construction"—and a small group of CCC men built the buildings between 1938 and 1940. According to Mr. O'Shia, who was "on the log crew" of about five CCC men, the crew "worked all the logs down with a big drawing knife made from grader blade steel" before the logs were hewn flat top and bottom, saddle-notched for corners, and incorporated into the structures. "A green car would visit from Denver every month,"

Mr. O'Shia recalled in 1995, bringing what the CCC enrollees referred to as "big shots" to check construction progress. "The old Scotsman doing the stone masonry would tell his helper to 'wash the rocks' when the green car would arrive." Work on this and other projects, including firefighting, went on year round. "We didn't have to work when it was 30 below zero," Mr. O'Shia reported.

At the end of the job, Lake George Ranger Station began four decades as headquarters of the old Lake George Ranger District. Ranger Becton seems to have been the first to occupy the residence, and Ranger Bill Zimmer was the last district ranger stationed there. In 1976, the district was incorporated into the South Park Ranger District headquartered at Fairplay. Since then, historic Lake George Ranger Station has been a work center staffed by two permanent and several seasonal Forest Service employees. The office, open five days a week from about mid-May through mid-October, provides information to the public. This information office, however, may be replaced by a new visitor center at Wilkerson Pass, about twelve miles to the west, within the next few years.

ACCESS: To get to historic Lake George Ranger Station from Colorado Springs, take U.S. Highway 24 west for 37 miles to the vicinity of Lake George. Just before entering the community, turn north onto Park County Road 94, Trail Creek Road. The ranger station is about 300 feet up this road on the right. The office is open from 8:00 a.m. to 4:30 p.m., Tuesday through Saturday, from about mid-May through mid-October. The log ranger station buildings are easily viewed from the sidewalk outside the office.

MESA LAKES RANGER STATION
Grand Mesa National Forest, Colorado
(1940)

There's been a ranger station at the Mesa Lakes on Grand Mesa, one of the world's largest flat top mountains, at least since 1905 when pioneer rangers like Bill Kreutzer and Jim Cayton rangered

Mesa Lakes Ranger Station, Grand Mesa National Forest, Colorado, in 1944. U.S. Forest Service photo.

western Colorado's forest reserves for the brand-new U.S. Forest Service. The first log cabin to house the station (photograph, page 6) was replaced by a second log dwelling in 1922. A barn had been added by 1934. Then, in response to growing need, the current Mesa Lakes Ranger Station log buildings were built by the Civilian Conservation Corps in 1940 and 1941.

When, in 1923, the Forest Service platted the Mesa Lakes for special use permit summer homes, the area's future was fixed. By the 1930s, the Mesa Lakes area had become a popular vacation spot. It was well known for its fishing. Among his many duties, the district ranger at Mesa Lakes administered this early recreation use. It fell to him to deal with disputes over fishing between members of the public who wanted to fish the lakes and private concerns that had purchased rights to private use of the lakes. The dispute ended when the Forest Service bought out the private fishing rights in the 1930s. Then, when the CCC completed the Lands End Road from the Gunnison Valley to Lands End Point on Grand Mesa and points

Mesa Lakes Ranger Station under construction on June 1, 1940: the new ranger's residence (left), the new office building (right) and the 1922 ranger station building (center) (above). Historic Mesa Lakes Ranger Station today (below). U.S. Forest Service photos.

beyond, the district's recreation activity and the ranger's recreation work increased dramatically. A new ranger station was needed.

In 1939, Forest Supervisor Ray Peck planned a new Mesa Lakes Ranger Station that would include a new office building, ranger's residence, and other needed facilities. The first two new buildings, the office and residence, were built next to the existing buildings that were left standing and used throughout construction. When the new log buildings were finished in the summer of 1940, the old ones were demolished.

Construction of the garage, which began with completion of the office and residence, wasn't finished by the time winter set in and deep snow threatened to halt work. But the CCC workers at Mesa Lakes built a frame structure around the garage project so work could continue through the winter, and the garage was finished by spring. A photograph taken in June, 1940, shows part of a similar frame structure erected around the nearly completed residence, and suggests the same tactic for winter construction work may have been used there.

No longer a district ranger's headquarters, historic Mesa Lakes Ranger Station—appearing today just about as it did during the 1940s—houses seasonal Forest Service employees and volunteers and provides visitor information during the summer.

ACCESS: To reach historic Mesa Lakes Ranger Station from the west, turn off Interstate Highway 70 onto Colorado Highway 65 about 20 miles east of Grand Junction, and follow that highway for about 27 miles east and then south onto the Grand Mesa and to the Mesa Lakes turnoff. From the east, turn off Interstate Highway 70 south of DeBeque and follow the DeBeque Cutoff Road for about 12 miles to its junction with Colorado Highway 65, then follow that highway south for about 16 more miles to the Mesa Lakes turnoff. From the Mesa Lakes turnoff, follow the road for less than 1/2 mile past the Jumbo Campground and around the northern shore of Sunset Lake to Mesa Lakes Ranger Station. The station is open six days a week from July 1 to mid-September, generally from 10:00 a.m. to 3:00 p.m. It's staffed by volunteers, and hours may vary.

Brush Creek Work Center office, Medicine Bow National Forest, Wyoming. U.S. Forest Service photo.

BRUSH CREEK RANGER STATION
Medicine Bow National Forest, Wyoming
(1941)

High in southern Wyoming, in a grassy clearing where the dense lodgepole pine forest of the Medicine Bow's western slope yields to the open sagebrush of the Kindt Basin, historic Brush Creek Ranger Station—three log structures built between 1937 and 1941 to replace an older ranger station—remains in service as Brush Creek Work and Visitor Center.

The first Brush Creek Ranger Station, built about a mile to the northwest in 1905 as headquarters of a Medicine Bow Forest Reserve district, was originally called Drinkhard Ranger Station. But the name didn't do much for the young Forest Service's image, and was changed to Brush Creek in 1914.

By the late 1930s, when the Medicine Bow National Forest needed a new Brush Creek Ranger Station, Depression-era public works agencies like the Civilian Conservation Corps were old hands at building ranger stations to Forest Service standard plans. Under foreman Herb Hohn, most of the heavy construction was completed between 1937 and 1940 by CCC crews from Ryan Park, a side camp of the Saratoga, Wyoming, CCC camp, during the warmer months of those years. These crews cut and peeled the lodgepole pine logs, excavated and laid the foundations, and finished all the rough construction. Local craftsmen were hired to do most of the skilled finish work. Fred Potter, an Encampment area rancher, for example, did all the stone cutting and masonry work. And Urban Schantz, a Finnish carpenter from Chicago, built the cabinets in the office and residence. The result was a pleasing group of three one-story log buildings—an office, a residence, and a garage—linked by cut stone sidewalks.

From this station, the district ranger administered the old Brush Creek Ranger District until it was combined with the old Bow River Ranger District to form the current Brush Creek Ranger District headquartered in Saratoga. When that happened, historic Brush Creek Ranger Station, to which several buildings were added over the years, became a work center.

Two rectangular, one-story, wood-frame storage buildings at today's Brush Creek Work Center, across Brush Creek from and not part of the historic ranger station, have an interesting history. Both were built in 1942 as part of the Ryan Park Prisoner of War Camp and moved to Brush Creek in the 1960s.

ACCESS: Historic Brush Creek Ranger Station is located on the east side of Wyoming Highway 130, the Snowy Range Scenic Byway, about 60 miles west of Laramie and about 9 miles east of Wyoming Highway 130's junction with Wyoming Highway 230. A visitor information center, open during the summer months, interprets the historic station and provides recreation information.

Chapter Three
SOUTHWESTERN REGION

The twelve national forests of the Southwestern Region, or Region 3, include twenty-one million acres in the states of Arizona and New Mexico—in general the cooler and wetter areas in an arid land. This vast land of high plateaus incised by deep canyons—the Grand Canyon among them—and surmounted by isolated mountain ranges that drops off to lower deserts, also punctuated by scattered mountain ranges, has many climates. These give rise to vegetation that varies from ocotillo and cactus in the desert plains to juniper, pine, aspen, and spruce-fir forests at higher elevations. These environments afford a variety of recreational opportunities. But the Southwestern Region's national forests are more than playgrounds. Most of the usable water available to Arizona and New Mexico

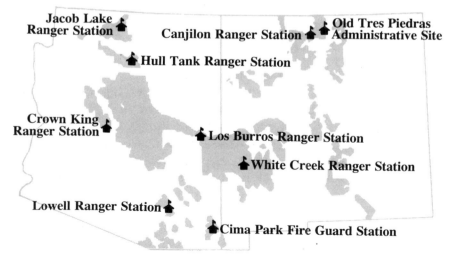

Nine historic U.S. Forest Service ranger stations of the Southwestern Region (Region 3) in Arizona and New Mexico.

91

flows from national forest lands that also supply timber for mills, grazing for cattle and sheep, habitat for wildlife, and other benefits. These lands are managed by a regional forester in Albuquerque, eleven forest supervisors, and about sixty district rangers.

In the early years, Southwestern Region forest rangers commonly used abandoned ranch and mine cabins for ranger stations. As did many of the administrative facilities later built by the Forest Service, these ranged from log cabins in the mountains to adobes on the edges of deserts, and included wood-frame structures. The nine described in this chapter are representative of perhaps dozens of Region 3's historic ranger stations.

HULL TANK RANGER STATION
Kaibab National Forest, Arizona
(1907)

The old Hull Tank Ranger Station, now called the Hull Cabin Historic District, lies in a small meadow surrounded by old-growth ponderosa pines about a mile south of the Grand Canyon. This site, commonly referred to simply as Hull Cabin, consists of a log cabin residence, a log storage cabin, a log barn, and a water tank. The appearance of this old ranger station has changed little since it was built as a sheep ranch in the 1880s, and none of the buildings has electricity, plumbing, or telephone service.

Hull Cabin was built by sheep ranchers Phillip and William Hull in 1888. That same year they constructed Hull Tank, an important source of water for early-day Grand Canyon visitors as well as sheep in this dry land. Both cabins are built of round ponderosa pine logs that are V-notched at the corners—a difficult construction technique. The barn, constructed in 1891 of massive hand-squared ponderosa pine logs joined by dovetailed notches at the corners, is of particular interest.

After his brother, Phillip, died in 1888, William Hull continued in the sheep business and occasionally guided tourists to the Grand Canyon. At that time, before completion of the Grand Canyon

Hull Tank Ranger Station on the old Tusayan National Forest—now part of the Kaibab National Forest, Arizona—about 1925. U.S. Forest Service photo.

Railroad in 1901 made Grand Canyon Village more accessible to visitors, nearby Grandview at the end of the Flagstaff stagecoach line was the center of Grand Canyon tourism and mining. By 1893, Hull had left the sheep ranching and tourism businesses to prospect in the Canyon's depths. The land surrounding Hull Cabin became part of the Grand Canyon Forest Reserve that same year, and Hull Cabin became U.S. Government property about the turn of the century. The Hull cabins became a U.S. Forest Service ranger station in 1907, and in 1908 the Forest Service built the third room onto the rear of Hull Cabin and, probably at the same time, the cabin's massive native stone chimney. In 1910, when national forest land south of the Grand Canyon became the Tusayan National Forest, Hull Tank Ranger Station became an important administrative site. Rangers there were responsible for a large area that, until establishment of Grand Canyon National Park under National Park Service administration in 1919, included the Grand Canyon. They lived and worked at the Hull Cabin station in the summer; in the winter, they moved to Anita, about 18 miles west along the Grand Canyon Railroad line.

George Reed, a Forest Service assistant ranger assigned to Hull Tank Ranger Station in 1907, remained there until 1919. In addition to carving his name and the date " '07" on the inside wall of the barn, Reed repaired the cabins and the road leading into the station, rebuilt Hull Tank and installed pipes from the tank to the cabins and the barn, and probably helped install a telephone line. Although remains of these improvements may be seen, the telephone and plumbing systems no longer work.

James E. Kinter followed George Reed as ranger at Hull Tank Ranger Station where he served until Arthur Gibson replaced him in 1924. Ranger Gibson replaced the barn roof with the corrugated metal that still bears his name and address in black paint for railroad delivery.

Bill Vogelsang assumed ranger duties from Arthur Gibson in 1930, and was ranger at Hull Tank Ranger Station when the Tusayan National Forest became the Tusayan Ranger District of the Kaibab National Forest in 1934. Ranger Vogelsang remained until 1940 when Hull Cabin ceased to be a ranger station. The Civilian Conservation Corps built a new station in Tusayan near the Grand Canyon National Park entrance between 1939 and 1942, and the ranger was transferred there.

A proposed agreement between the National Park Service and the Forest Service to transfer Hull Cabin ownership from the Kaibab National Forest to Grand Canyon National Park was never acted on by Congress. Had the transfer occurred, the Park Service probably would have destroyed Hull Cabin along with the other old structures in the Grandview area. But Hull Cabin remained to house Forest Service seasonal crews.

The Forest Service ensured that the former Hull Tank Ranger Station would stand for another century when it stabilized the cabins and barn in 1989 and 1990. Using original building methods and materials, Forest Service employees from throughout the West helped replace rotten logs in the main cabin and barn, reroofed both cabins, rebuilt the cracked chimney on the main cabin, and replaced clapboards on the barn gable end.

The Hull cabins and barn, for thirty-three years a Forest Service ranger station, remain to remind visitors of earlier eras on the South Rim. Visitors are welcome at Hull Cabin, and are expected to treat this historic site with respect.

ACCESS: Hull Cabin Historic District is located about 15 miles east of the town of Tusayan and Grand Canyon Village, and may be reached from either. Just south of Tusayan, turn east off Arizona Highway 64/180 onto Forest Road 302 and follow it eastward. South of historic Grandview Lookout, Forest Road 302 becomes Forest Road 307. Follow the signs on Forest Road 307 for about 1 1/2 miles to Hull Cabin. From Grand Canyon Village, drive eastward along the South Rim of the Grand Canyon on Arizona Highway 64 for about 15 miles. About 2 miles east of the Grandview Point turnoff, turn right at the Arizona Trail sign and follow Forest Road 310 onto the Kaibab National Forest. Pass historic Grandview Lookout after a little more than 1 mile, turn left onto Forest Road 307, and follow the signs for about 1 1/2 miles to Hull Cabin. Overnight camping is prohibited within 1/4 mile of the Hull Cabin Historic District.

JACOB LAKE RANGER STATION
Kaibab National Forest, Arizona
(1910)

About forty miles north of the Grand Canyon, high on the Kaibab Plateau, historic Jacob Lake Ranger Station is the only one of six early-day Kaibab National Forest ranger stations north of the Canyon that remains. Built by the Forest Service in 1910, this cabin and barn face the meadow surrounding Jacob Lake, a natural pool converted into a cattle tank. Decades before, Mormon pioneer Jacob Hamblin lent his name to that small but permanent source of water that attracted cattlemen and others to the area.

Jacob Lake Ranger Station was built about twenty years after cattle grazing and lumbering, which began on the Kaibab Plateau about 1890, demonstrated the need for government administration

Ranger Will Mace (leaning against porch post) and members of his family on the front porch of the Jacob Lake Ranger Station cabin under construction in 1910 (above), and historic Jacob Lake Ranger Station after restoration in 1991 (below). U.S. Forest Service photos.

of the region's natural resources. It took most of those twenty years for the government to sort out that administration. By 1908, the forest reserve land north of the Grand Canyon had been renamed the Kaibab National Forest (the part adjacent to the North Rim of the Grand Canyon was included in Grand Canyon National Park under National Park Service administration in 1919), and the new Forest Service set about building ranger stations from which to administer these lands.

The memoirs of early-day Ranger Will Mace, written as a 1940 letter to Gifford Pinchot, tell the Jacob Lake Ranger Station story best. Mace, a son of Kanab, Utah, pioneers and a high school graduate in his early twenties, sought opportunity in Pinchot's new Forest Service.

> In preparation for the appointment which I hoped to receive, I quit my job at the sawmill at Jacobs Lake and early in January 1909 had registered at the Utah State Agricultural College (in Logan) for a short course (8 weeks) in Forestry which was being given for the first time that year. . . . The course was conducted by Julian E. Rothery, Yale Forest School graduate of 1908, . . . who did a very creditable job of introducing the class to the rudiments of Technical Forestry and Surveys. For me this proved to be very valuable training.

Mace was appointed an assistant forest ranger on the Kaibab National Forest on April 1, 1909, and in addition to his Jacob Lake Ranger District assignment, "assisted on several pioneering projects" in range and timber management during the next two and one-half years. He recalled building Jacob Lake Ranger Station.

> On July 1, 1910, it was necessary for me to again take up the duties of my Ranger District. Among other things I supervised the finishing work on the two room cabin at Jacob's Lake, begun the year before, and built a combined stable and hay barn.

As built, both the cabin and barn were rectangular wood-frame structures with cedar-shake-covered gable roofs and board-and-batten siding. The ponderosa pine lumber used in the structures was cut locally and milled at the now-dismantled Jacob Lake Sawmill. Although the cabin had a stone and mortar foundation, the barn

97

survived without a foundation until one was added in the 1950s. Former Ranger Mace noted with pride in his 1940 letter that "These buildings are still being used by the Rangers." Indeed, a major remodeling in 1941, during which the cabin's front porch was enclosed to add a bedroom and a bathroom, and improvements and maintenance throughout the years ensured Jacob Lake Ranger Station's active use through the 1980s.

Today, with the cabin restored to its original appearance, the cabin and barn at Jacob Lake—the only surviving ranger station of board-and-batten construction on the entire Kaibab National Forest—look just about as they did in 1910. Plans call for interpretation at the cabin to evoke the life and work of Ranger Mace and other Forest Service rangers of the time.

And what a time it was! During the winter and spring of 1910, for example, the yet-to-be-finished Jacob Lake Ranger Station was base camp for a six-month survey of the Kaibab National Forest's timber resources conducted by Ranger Mace and several other rangers. For two of those months, the survey team worked on snowshoes. Supplies were hauled in over deep snow by dog sleds—dog sleds that Ranger Mace recalled "made dandy toboggans for Sunday afternoon recreation." The data from this pioneering project, the first ever in the Southwestern Region, provided a baseline for future forest management.

A more famous management effort, the hunting of predators—especially mountain lions—to protect the Kaibab Plateau's deer, followed President Theodore Roosevelt's proclamation of the Grand Canyon National Game Preserve in 1906. Since the preserve was within the forest reserve—soon redesignated a national forest, the Forest Service was charged to protect the deer and hired a game warden to systematically hunt predators. Warden Jim "Uncle Jimmy" Owen, the legendary "cougar killer of the Kaibab," carried out his duties relentlessly—sometimes from Jacob Lake Ranger Station. Ranger Mace recalled having . . .

> . . . had one or two exciting hunts with Uncle Jim Owens (sic) and his pack of hounds. Whenever he had a trip to make after mountain lions into rough, isolated territory along the rim of the Grand Canyon, if other duties permitted one of the Ranger force

Uncle Jimmy Owen, the "cougar killer of the Kaibab," at Jacob Lake in 1920. U.S. Forest Service photo.

was assigned to accompany him. I was always pleased when I drew one of these assignments—as were the other boys—because we like (sic) the excitement.

Uncle Jimmy guided former President Roosevelt and his sons on a month-long cougar hunting trip on the Kaibab National Forest during the summer of 1913. But eliminating cougars proved disastrous. Without predators, the deer population grew rapidly and forage became scarce. By 1920, Kaibab Plateau deer began to starve in massive numbers, and the Forest Service tried several solutions to the problem including a failed 1924 attempt to drive the herd across the Grand Canyon to the South Rim. Finally, the Kaibab National Forest got Supreme Court permission to allow deer hunting within the game preserve. Today, deer hunting is one of the most popular recreational activities on the Kaibab Plateau.

ACCESS: Jacob Lake Ranger Station is located about 1 1/2 miles southwest of the Kaibab Plateau Visitor Center. To get there, take Arizona Highway 67 south from its intersection with U.S. Highway 89A for about 1/2 mile. Turn right (west) onto Forest Road 461, following the "camping" sign, and drive for about 1/2 mile, then turn left (south) onto Forest Road 282 to a private campground. Jacob Lake Ranger Station, identified by a small interpretive sign atop a stone pedestal, is located on the east side of this road across from the campground and Jacob Lake, a stock watering tank. Until the cabin is opened for interpretation, visitors may view the outside of the cabin and peek in the windows.

LOS BURROS RANGER STATION
Apache-Sitgreaves National Forests, Arizona
(1910)

The two board-and-batten buildings of historic Los Burros Ranger Station, built on the Sitgreaves National Forest of east-central Arizona in 1910, are the oldest existing Forest Service structures on the Apache-Sitgreaves National Forests. Set amid the cinder cones of the White Mountains volcanic field at about 7,800 feet above sea level, the Los Burros station—situated among oak, ponderosa pine, and aspen, and overlooking a meadow—is surrounded by a ponderosa pine forest.

Protection of that valuable virgin ponderosa pine forest from wildfires was Los Burros Ranger Station's original purpose. The Sitgreaves National Forest, especially its Mogollon Rim country, has one of the highest known incidences of lightning strikes. For at least a third of the year, the fire threat is real and continuous. To counter this threat, the Forest Service posted fire guards around the forest. One of these guards was assigned to Lake Mountain, at 8,500 feet the highest vantage point in the area. About a mile to the south, the Forest Service built Los Burros Ranger Station—a combination house-office and barn for this guard and his horses—along the only road through the area. A perennial spring at Los Burros provided a dependable water supply, and the nearby meadow provided pastures for the large number of saddle and pack animals that would transport a large fire crew to the area. The station was connected to the district ranger's office, twenty-five miles away in Show Low, by a bare-wire telephone line strung from insulators nailed to trees.

During the fire season, the Los Burros fire guard rode to the top of Lake Mountain—a lookout tower was built there in 1926 or 1927—where he could watch over a vast area. When he sighted a smoke, he could return to Los Burros, call the ranger in Show Low, and order up firefighters who would ride to the fire and put it out with hand tools.

Use of Los Burros Ranger Station expanded in 1918 when the Sitgreaves National Forest sold 250 million board-feet of timber to Cady Lumber Company in nearby McNary. For the next two or three decades, the station housed the Forest Service timber sale administrator as well as the fire guard. The role of the sale administrator—indeed, the primary role of the Forest Service during this period—was to ensure proper harvesting of the timber resource. Although the science of silviculture was young, sustained-yield timber production was at the core of Forest Service philosophy. Foresters camped at Los Burros while they measured timber stands and monitored timber cutting on this sale, one of the

101

Sitgreaves's first large timber sales. As a result of this sale, McNary grew as a logging and mill town.

Los Burros Ranger Station was also in the midst of ranching country. In addition to Forest Service personnel and loggers, cowboys on roundups and sheepherders driving sheep used the spring and meadow. Indeed, early sheepherders who camped there probably named Los Burros for stray pack animals that roamed the area. Los Burros remained an overnight stop on the main sheep driveway through this country until the 1970s.

Farming also had its day near Los Burros. As reported by Jo Baeza in the February 13, 1987, issue of the *NavApache Independent*, Don Hansen of Lakeview, Arizona, leased thirty acres of national forest land just south of Los Burros at Reservation Flat for one dollar per acre per year. Year after year, between 1925 and 1937, Hansen raised "bumper crops of potatoes, carrots, turnips, onions and peas" in that "black, loamy soil" for sale in the Salt River Valley. He also raised so much wheat, oats, and sometimes barley that "the sacker couldn't sack the grain fast enough when it was being thrashed." One snowy night in 1930. Hansen took refuge in the ranger station barn. Awakened in the middle of the night by a great weight on his legs, he sat up and startled an "animal" that, as he put it, "pulled freight—jumped out the window." Hansen never knew for sure what slept with him in the Los Burros barn that night, but some time later he saw a large bear amble up the trail from the ranger station spring.

Don Hansen also recalled the Rogers family from Snowflake, Arizona, that lived at Los Burros Ranger Station. Mr. Rogers, apparently a Forest Service employee, had a number of good-looking, musically-inclined daughters. As Hansen put it: "All the boys in the country knew where Los Burros was."

When the large timber sale ended in the late 1940s or early 1950s, Los Burros Ranger Station fell into disuse. Both the two-room house-office and the barn are of one-inch by twelve-inch board and one-inch by four-inch batten construction. The outside walls, which appear to have been red, are now painted a cream color. Although the house-office roof is now covered by green asphalt shingles, it probably had a wood-shingle roof like the barn retains. The spring, developed with a concrete liner and cap covered with native stone, blends into the landscape. Despite a few decades of neglect, Los Burros Ranger Station remains in fair condition and provides a window on early Forest Service history.

ACCESS: Normally accessible from mid-April through October, historic Los Burros Ranger Station is located just south of Forest Road 224 (the Vernon-McNary Road) about 8 miles northeast of McNary. McNary is on Arizona Highway 260 about 20 miles southeast of Show Low and about 40 miles west of Springerville. Los Burros Campground, a national forest campground with tables and fire rings at eight sites but no water or trash collection, is adjacent to the historic ranger station. The historic Lake Mountain fire lookout tower is nearby.

OLD TRES PIEDRAS ADMINISTRATIVE SITE
Carson National Forest, New Mexico
(1913)

Aldo Leopold, a young forester from Yale who served as Carson National Forest supervisor from May 1911 until March 1913 and later became a leading American conservationist, was twenty-five years old and in love when he designed and built the house that dominates the Forest Service administrative site at Tres Piedras in northern New Mexico.

The two-story, nine-room, wood-frame "Leopold House" is the oldest structure in the Old Tres Piedras Administrative Site Historic District, and the only one associated with Leopold's relatively short term as Carson National Forest supervisor. Leopold drew the house plan in 1911, saw to it that "six-hundred-and-fifty large round silver dollars, coin of the realm" were appropriated for its construction, and built it in 1912 as a labor of love for his bride-to-be, Miss Estella Bergere of Santa Fe, with whom he had discussed the project. According to Leopold biographer Curt Meine:

Old Tres Piedras Administrative Site, Carson National Forest, New Mexico, showing Ford pickup acquired in 1919 (above) and in 1946 (below). U.S. Forest Service photos.

Deputy Forest Supervisor Aldo Leopold, Forest Assistant Ira T. Yarnell, and Forest Supervisor Harry C. Hall at Carson National Forest headquarters in Tres Piedras, New Mexico, in 1911, before Leopold built "Mia Casita" in 1912. Photo by Raymond Marsh courtesy of Forest History Society.

He planned the house to face east over the thirty-mile-wide valley to the snow-topped Sangre de Cristos. It had to be done just right—simple, elegant, by necessity small, and set amid the granite boulders and pinon pines of Tres Piedras. And, of highest priority, it had to have a great fireplace.

The Leopolds called the house "Mia Casita." They lived there from their marriage in October 1912 until Leopold's April 1913 trip to the Jicarilla Ranger District during which he contracted an illness that resulted in two years of sick leave followed by assignment to a Forest Service staff position in Albuquerque.

Changed only slightly from the plans Leopold drew in 1911, the house has been a Forest Service family residence from Leopold's day to our own. The current residents are the Tres Piedras district ranger and his family.

The other historic structures at the Old Tres Piedras Administrative Site—which began as a forest supervisor's headquarters
rather than as a ranger station—are the office, barn, and several
small outbuildings. The office, built about the same time as the
house but six miles west as Cow Creek Ranger Station, was moved
to Tres Piedras about 1917. Ranger Elliot Barker, who had joined
the Carson National Forest force in 1912, was Leopold's ranger at
Cow Creek and later a distinguished conservationist. In 1950, as
New Mexico's state game warden, Barker donated a bear cub
rescued from a fire on the Lincoln National Forest to the Forest
Service. Domiciled at the National Zoo in Washington, D.C.,
Smokey Bear became America's living symbol of forest fire prevention. Originally a single-room cabin with a pot-bellied stove,
the office has been partitioned into two rooms and used for a variety
of purposes. The barn, just north of the house, was built in 1931
as a shelter for livestock and their feed. As a group, these buildings
evoke a bygone era when administration of a national forest or a
component ranger district was essentially a one-man operation.

The two years Aldo Leopold spend at Tres Piedras were basic
to the insights reflected in his "land ethic" and his many contributions to conservation. There, in 1913, he first raised the possibility
of setting aside remote areas of wilderness as part of the national
heritage and for scientific study. As a result of his efforts, those
parts of the Gila National Forest now known as the Gila Wilderness
became the nation's first "wilderness" in 1924 and the forerunner
of today's more than 100-million-acre National Wilderness Preservation System. Leopold left the Forest Service in 1928 to continue
a professional career, in Meine's words, "at the cutting edge of
conservation activity and environmental thought." His 1949 book,
A Sand County Almanac, is a classic of environmental literature.

Although the Old Tres Piedras Administrative Site Historic
District is part of a working administrative center, visitors are
welcome at this living memorial to the early days of the Forest
Service and the development of America's conservation ethic.

*ACCESS: The Old Tres Piedras Administrative Site is located at
the base of the central of the Tres Piedras—three large granite
extrusions that dominate the surrounding countryside—immedi-*

106

ately north of the town of Tres Piedras. Turn west off U.S. Highway
282 about one-eighth mile north of Tres Piedras and follow the road
about one-eighth mile to the site. View the historic Leopold House
and the office and barn from a distance that respects the privacy of
the residents.

WHITE CREEK RANGER STATION
Gila National Forest, New Mexico
(1933)

When, in May of 1909, a forest guard—later a ranger—named
Henry Woodrow "made camp where the old White Creek Station
was finally built," he couldn't have known he'd located his summer
base for the next thirty-three years. The ranger stations eventually
built there, at the confluence of White Creek and the West Fork of
the Gila River in the Mogollon Mountains of southwestern New
Mexico, served as summer headquarters for the Gila National
Forest's old McKenna Park Ranger District—later the Wilderness
Ranger District in the Gila Wilderness—for fifty-nine years. And
historic White Creek Ranger Station, now called White Creek
Cabin, continues in occasional use by Forest Service wilderness
rangers and trail crews.

Ranger Woodrow rangered the verdant McKenna Park high
country—put out fires, built trails, and "looked after grazing"—
from his White Creek camp for four summers. Then, "In the fall"
of 1912, he wrote in his 1943 memoir, "$75.00 was appropriated
to build a log cabin. . . . I hired a man to help and we put up a log
cabin and covered it with shakes or boards split from a pine tree."
With a summer station at White Creek, he wintered at Gila Ranger
Station about twenty miles to the southwest. This winter headquarters brought romance into his thirty-second year of life. As he wrote
in 1943:

> There happened to be a widow on this part of the District with a
> grazing permit on the Forest and a ranch near the Gila Station.
> So I married her on October 14, 1912.

107

White Creek Ranger Station, Gila National Forest, New Mexico, in 1939. U.S. Forest Service photo.

Then he philosophized about the practicality of forest rangers marrying ranching widows, and revealed his great faith in matrimony:

> Later I heard of Rangers on other Forests and Districts having quite a bit of trouble with widow permittees. . . . I would suggest that the Forest put a single man for Ranger there and probably he would marry her and stop all the trouble.

Ranger Woodrow spent every summer at White Creek Ranger Station until his 1942 retirement. The station got its first telephone line in 1914—"some small insulated copper wire . . . on limbs of trees . . . but not very satisfactory"—and a better one in 1915. Before and during World War I, he and a man or two built firemen's cabins at Little Creek and Snow Park, a lookout of sorts on Granite Peak, miles of trail, and fences to improve White Creek pastures. In 1923, the New Mexico Department of Game and Fish started a small fish hatchery, probably to propagate the Gila trout.

In 1933, the cabin that would replace Ranger Woodrow's 1912 log cabin as White Creek Ranger Station was built, not by the Forest Service, but by I.V. Lash for the State of New Mexico as part of that fish hatchery. The logs were cut locally, and other materials were packed in from Willow Creek by Vance Hawkins. In 1937, according to Ranger Woodrow:

> . . . the Forest Service took over the fish hatchery buildings at the mouth of White Creek, so we abandoned the old White Creek Station where I had spent 28 summers. We moved down and repaired the house and fences, and constructed 4 1/2 miles of telephone line to the new station from Willow Creek and Silver City line. We repaired the bunk house and tool house, and built 1/2 mile of new pasture fence around the station.

In 1938, a small Civilian Conservation Corps crew from the Willow Creek CCC camp installed a water system that provided hot water and a bathroom with a flush toilet and a shower for the White Creek Ranger Station cabin. They also cut logs for a new barn that they completed in 1939 and remains.

As remote as it was, White Creek Ranger Station had its share of notable visitors. In 1922, Aldo Leopold, then in the Albuquerque regional office and surveying the Gila National Forest with Forest Supervisor Fred Winn as a possible wilderness area, showed up to help Ranger Woodrow and others fight a string of bad fires. Two years later, as a result of Leopold's leadership, the district became part of the Gila Wilderness Area, the first area in the National Forest System and the nation to be so designated. In the 1950s, after Ranger Woodrow's time, Senator Clinton P. Anderson of Arizona, who served President Harry S. Truman as Secretary of Agriculture from 1945 to 1948, was an occasional visitor. As retired Ranger Chuck Hill, the last district ranger to use White Creek Ranger Station as a summer headquarters, wrote in 1993:

> . . . Senator Anderson did the Forest Service a lot of favors. The Forest Service responded by fixing up White Creek Cabin so he could visit when possible and enjoy his comforts . . . even scratched in a tractor trail from Willow Creek. . . . The senator rode in to White Creek on a trailer behind a little Ford tractor.

Forest Service wilderness rangers and trail crews use historic White Creek Ranger Station, now called White Creek Cabin, to support Gila Wilderness operations. Photo by Dick Spray.

By the time Ranger Hill arrived in 1963, Senator Anderson's visits were a thing of the past. And, with inclusion of the Gila Wilderness in the National Wilderness Preservation System created by the Wilderness Act of 1964, the forest supervisor had Ranger Hill

> . . . begin eliminating the two-rut tractor trail in a passive manner. When a tree fell across it, we were to cut out enough to open only one track. In the event of an earth or rockslide, the same practice was followed—we dug out enough to open a single track.

In deference to the area's wilderness status, such improvements as the ranger station's water and electrical systems also were removed.

"I suppose that station was just about right back in Henry Woodrow's day, when district activities actually centered there," retired Ranger Hill reflected in 1993. "By the time I got there it had become somewhat of a white elephant." It was expensive and difficult to maintain. At the end, when helicopter-borne fire crews stationed outside the wilderness replaced horseback firefighters, the fire control funds that had kept White Creek Ranger Station in operation were cut off. In 1968, the Wilderness Ranger District headquarters was moved to Gila Center. With this move, the Forest Service dismantled many of the corrals and pasture fences at White Creek Ranger Station and other administrative camps in the wilderness.

Wilderness Ranger District is now administered from Mimbres Ranger Station.

ACCESS: The cabin and barn that comprise historic White Creek Ranger Station, deep within the 557,873-acre Gila Wilderness, may be reached only on foot or by horseback. The better accesses are from Gila Center, Sandy Point, and Willow Creek. The most direct routes from the Gila Center vicinity are Forest Trail 151 up the West Fork of the Gila River, about 19 miles, and Forest Trail 164 through Woodland and Lilley parks, about 22 miles. From the Gila Center area, Forest Trail 151 begins at the end of New Mexico Highway 15, about 37 miles north of Silver City and about 1 1/2 miles past the Gila Visitor Center near Scorpion Campground. The first 1/2 mile of the trail passes through the northeastern corner of Gila Cliff Dwellings National Monument before entering the Gila Wilderness. Forest Trail 164 begins at a trailhead along New Mexico Highway 15 midway between Gila Visitor Center and Gila Cliff Dwellings National Monument. A good downhill route of about 20 miles begins at Sandy Point, about 8 miles east of Mogollon on New Mexico Highway 159, and proceeds about 11 miles along Forest Trail 182 to the historic Mogollon Baldy Peak fire lookout and cabin, on about another mile to Forest Trail 152, and then about 8 miles on Forest Trail 152 via Snow Park to Forest Trail 151, and about another mile on that trail to White Creek Cabin. From the Willow Creek area, about 17 miles east of Mogollon on New Mexico Highway 159, Forest Trail 151 proceeds about 15 miles via Iron Creek Lake, Iron Creek, Cooper Canyon, Turkey-feather Pass, the West Fork of the Gila River, and Cub Mesa to White Creek Cabin. Use a Gila Wilderness map to plan this trip carefully, and obtain current information concerning trail conditions, water availability, etc., at the Gila Visitor Center or at ranger stations in Mimbres, Silver City, Glenwood, or Reserve.

Lowell Ranger Station, Coronado National Forest, Arizona, in 1936. U.S. Forest Service photo.

LOWELL RANGER STATION
Coronado National Forest, Arizona
(1934)

About a dozen miles northeast of downtown Tucson, at the base of the Santa Catalina Mountains, historic Lowell Ranger Station's three Pueblo-style buildings—an office, a residence, and a garage-shop—look right at home in their Sonoran Desert setting. That, of course, is by design. The buildings at Lowell Ranger Station, built by the Civilian Conservation Corps between 1934 and 1937 to standard Southwestern Region plans, were carefully fit into this environment.

All three historic Pueblo-style buildings at Lowell Ranger Station are built of stucco-covered adobe brick, and have flat roofs with decorative parapets. Design features of the office and residence include heavy wooden doors and covered porches with ceilings of thin wooden slats and peeled log beams. The residence had a patio. A hay window, which now has a wooden cover, suggests the garage-shop also served as a barn. These buildings are widely spaced on grounds heavily vegetated by saguaro cacti and other natural desert plants. An unpaved, rock-lined road leads between the buildings, passing the barn-garage and residence before circling in front of the office. In the original layout, the main road ran south of the office, making it the first building encountered by visitors.

Back when it was headquarters of the Santa Catalina Ranger District, Lowell Ranger Station was the Coronado National Forest's main visitor contact point for the popular Sabino Canyon Recreation Area until the Sabino Canyon Visitor Center was built in the early 1960s. No longer a ranger district headquarters, the historic Lowell Ranger Station adobes now house Forest Service volunteers and others—including Arizona Boys Ranch enrollees—who work on the Coronado National Forest.

Little changed over the years, historic Lowell Ranger Station retains the feel of the desert southwest in the 1930s and 1940s. In the future, the Forest Service hopes to develop the residence as a CCC museum.

ACCESS: Historic Lowell Ranger Station is located northeast of Tucson immediately behind the Sabino Canyon Visitor Center, where parking is available. To reach the visitor center from Tucson, turn north off Tanque Verde Road onto Sabino Canyon Road and drive north about 4 miles to the Sabino Canyon Visitor Center entrance on the right. The visitor center is open year-round from 8:00 a.m. to 4:30 p.m., Monday through Friday, and from 8:30 a.m. to 4:30 p.m., Saturdays and Sundays. To help respect the privacy of the old Lowell Ranger Station's residents, it's a good idea to make an appointment through the visitor center, telephone (520) 749-8700.

Cima Park Fire Guard Station, Coronado National Forest, in 1947. U.S. Forest Service photo.

CIMA PARK FIRE GUARD STATION
Coronado National Forest, Arizona
(1934)

High in the Chiricahua Mountains of southeastern Arizona, in a narrow canyon in the Chiricahua Wilderness, the log buildings of Cima Park Fire Guard Station look just about as they did when the Civilian Conservation Corps finished building them in 1934. At an elevation of about 9,000 feet, these buildings—a log cabin, barn, and shed, and a wood-frame outhouse—blend well with the surrounding forest of spruce and fir.

As its name suggests, Cima Park Fire Guard Station was originally a headquarters camp at which Forest Service fire crews were stationed during the dry, windy spring and early summer months during which wildfires are most likely in the high, remote Chiricahua Mountains. Telephone lines from fire lookouts atop peaks throughout the Chiricahuas were connected to the Cima Park station, where the cabin was home to a dispatcher who doubled as

114

a cook and to a seasonal fire crew. The barn housed the horses and mules needed to transport firefighters and their equipment to and from fires. When a lookout spotted a fire, he telephoned the Cima Park dispatcher who sent men to fight it.

Although modern fire fighting techniques that employ aircraft and sophisticated detection and communications systems have made the Cima Park station's headquarters role a thing of the past, it continues in use by wilderness rangers and firefighters.

ACCESS: Located in the Chiricahua Wilderness, Cima Park Fire Guard Station is accessible only by foot or horseback. The trail to Cima Park begins at historic Rustler Park Fire Guard Station, built by the CCC in 1935 as a main supply headquarters for fire lookouts and fire guard stations—including Cima Park Fire Guard Station— in the Chiricahua Mountains and still used as a base for a small crew of firefighters. Rustler Park is 120 miles east of Tucson, and 18 miles west of the small town of Portal. From Tucson, take Interstate Highway 10 east for 81 miles to Willcox. Drive southeast from Willcox on Arizona Highway 186 for about 33 miles to its junction with Arizona Highway 181. Turn left onto Arizona Highway 181 toward Chiricahua National Monument, and drive 3 miles, then turn right onto Forest Road 42, a gravel road. From Douglas, take U.S. Highway 80 two miles west to U.S. Highway 191. Go north on U.S. Highway 191 for 35 miles to Sunizona. Turn east (right) onto Arizona Highway 181 and drive east, then north, for approxi- mately 28 miles, staying on the paved road, to Forest Road 42. Ascend the Chiricahuas up Pinery Canyon on Forest Road 42 for 12 miles to Forest Road 42D, signed for Rustler Park. Turn right at Onion Saddle, and drive about 2 1/2 miles to Rustler Park. Forest Roads 42 and 42D are gravel roads suitable for passenger vehicles. Open from April through November, they are not plowed and are usually closed following early or late season snowstorms. These roads are rough and dusty, and may be muddy and slick after a rain. From Rustler Park, follow Forest Trail 270, the Crest Trail, south for about 3 miles to Cima Saddle, then follow Forest Trail 248 east for about 1/4 mile to historic Cima Park Fire Guard Station. Before using this trail, call the Douglas Ranger District at (602) 364-3468 for a report on current trail conditions and weather.

Crown King Ranger Station, Prescott National Forest, Arizona, in the late 1930s. U.S. Forest Service photo.

CROWN KING RANGER STATION
Prescott National Forest, Arizona
(1934)

The one-time Crown King Ranger Station, now Crown King Work Center, was headquarters of the first ranger district in Arizona for thirty district rangers over almost eight decades. Deep in the rugged Bradshaw Mountains southeast of Prescott, both the ranger station and community were named for the 1880s Crown King Mine, the first mine in the area. By the time the Prescott Forest Reserve was established in 1898 and what became Crown King Ranger District was set up in 1902, much of the area's ponderosa pine forest had been cut to provide mine timbers and fuelwood for the burgeoning mines. Remaining pines and junipers combined with dense manzanita and oak brush to cover the otherwise denuded surrounding hills.

The early ranger's life at Crown King was just about as rough as the country he worked. Isolation and improvisation ruled. Except for the railroad that passed through town and an almost impassable wagon road from Prescott, travel was by horse on trails or through brush thickets that made leather boots and chaps vital. At first, Crown King Ranger Station occupied rental quarters. Then, sometime between 1914 and 1916, Ranger Ed Ancona—a 1912 Penn State forestry graduate—actually bought his own ranger station! As he related this "little wrinkle of the early days when you had to make do" years later:

> There was four feet of snow in Crown King . . ., and the woman who owned the house that we were [renting] as a Ranger Station [said] I had to buy [the house], or else. So I . . . stayed there and bought the Ranger Station. But we found that the land under [the house] was National Forest land, so we staked out a mining claim [in Ranger Ancona's mother's name to avoid violating the Forest Service rule against forest officers owning mining claims on national forest land] to cover it.

Ranger Ancona "got into trouble only once" in his two years at Crown King, a result of his "make do" effort to build a barn for his ranger station.

> I stole a lot of telephone . . . poles [that] hadn't been used for about ten or fifteen years, but some company still . . . claimed them. . . . I went out and chopped down a number of those four by six redwood poles of an abandoned telephone line that used to run from Prescott to Phoenix. The whole country was pulling in these poles. . . . I built a nice barn out of them, but they found out afterwards that I had stolen those telephone poles. I suppose there's a mark somewhere on my record that I was a high-grader.

Ranger Ancona's work from his Crown King Ranger Station emphasized small sales of mining timbers and a lot of range management and trail work. He got on well with "tough fellows, those cattlemen and cowboys of that day" with whom he rode on roundups and even got to do some volunteer trail work. "That was something to do; get a cowboy off his horse and get him to take a pick and shovel and help build a trail." The only time Ranger Ancona got "really sore" at his Crown King ranger job was the time the community had an ice cream party at a neighbor's house. As he told it:

117

We were just about ready to serve the ice cream when a fellow rode up outside and knocked on the door and he said, "Is the Ranger here?" I scringed down, and somebody said, "He's over there." He says, "Well, you've got a fire up on the ridge above here. The lightning just struck a big pine. I can see it from my place; it's on the second or third ridge over." The call of duty was stronger than that of the ice cream; I had to leave. I never got any of that ice cream, and that's one of the big regrets of my life. I went out and sat by that burning tree all night while my friends were eating . . . the only ice cream that was in Crown King while I was there that two years. But I had put the fire out, by golly.

The end of the "make do" era at Crown King Ranger Station began in 1917, not long after Ranger Ancona's transfer to New Mexico, when the Forest Service bought a small house and barn for $1,500 that served as the ranger station for almost two decades. The ranger's life got better still when, in the mid 1930s, the Civilian Conservation Corps replaced the small house and barn with the distinctive Bungalow-style office, residence, and barn that comprise historic Crown King Ranger Station today. Sixteen district rangers administered the 161,950-acre Crown King Ranger District from these buildings between the time they were built and 1979 when the district was combined with another to form the Bradshaw Ranger District.

Nestled among ponderosa pines, these impressive Depression-era structures feature timbered gables and a unique flat stone exterior. Except for minor changes, they appear just about as they did when built. Still in use, they house a year-round Forest Service staff of three or four augmented to eight or ten during the summer.

ACCESS: The community of Crown King and historic Crown King Ranger Station are reached by leaving Interstate Highway 17 at Exit 259, about 30 miles south of Prescott and 60 miles north of Phoenix, and traveling west on the Bloody Basin Road for about 25 miles. This maintained gravel and dirt road that should pose no problem to passenger cars passes through the "almost ghost towns" of Cordes and Cleator. Depending on the season, about 60 to 100 people live in and around Crown King which has a store, two restaurants with bars, and a volunteer fire department—but few facilities for visitors seeking city comforts. There are three developed campgrounds at Horse Thief Basin, 7 miles south of Crown King.

Canjilon Ranger Station, Carson National Forest, New Mexico. Photo by William A. Westbury.

CANJILON RANGER STATION
Carson National Forest, New Mexico
(1935)

The adobe buildings of Canjilon Ranger Station—an office, a residence, and a garage-shop building—look just about as they did when Ranger Ed Grosbeck supervised their construction in 1935. About the only change is that the ends of the *vigas,* the roof support timbers, have been cut off and plastered over. These historic adobes remain in use on the Canjilon Ranger District.

ACCESS: Canjilon Ranger Station is located in the town of Canjilon, 3 miles east of the junction of U.S. Highway 84 and New Mexico Highway 110.

Twelve historic U.S. Forest Service ranger stations in the Intermountain Region (Region 4) that includes Utah, Nevada, southern Idaho, southwestern Wyoming, and a part of eastern California.

Chapter Four
INTERMOUNTAIN REGION

The eighteen national forests of the Intermountain Region, Region 4, include thirty-one million acres in Utah, Nevada, southern Idaho, southwestern Wyoming, and eastern California. These national forests, reaching from the towering Rocky Mountains in the east and north to the abrupt escarpment of the Sierra Nevada in the west, center on the numerous north-south trending mountain ranges separated by broad arid valleys that geologists call the Basin and Range Province and geographers often tag the Intermountain West.

Above the sagebrush-scrub vegetation of the basins, Region 4 national forests are characterized by juniper-pinyon woodlands at middle elevations and often-sparse coniferous forests higher in the mountains. Almost a third of the region's national forest acreage is commercial timberlands, and more than half of the cattle and sheep permitted to graze on national forests graze this region's rangelands. The region is rich in mineral resources. Recreation opportunities abound. And all the region's lands are valuable as watershed, wildlife habitat, and scenery. A regional forester in Ogden, thirteen forest supervisors, and about seventy district rangers manage these lands.

A dozen of the Intermountain Region's historic ranger and guard stations are profiled in this chapter.

"Rosie's Cabin" at the first Blackrock Ranger Station, now the Rosencrans Cabin Historic District, Bridger-Teton National Forest, Wyoming. U.S. Forest Service photo.

BLACKROCK RANGER STATION
Bridger-Teton National Forest, Wyoming
(1904)

An amazing Austrian immigrant named Rudolph Rosencrans, who became the Buffalo Ranger District's first ranger and built the first Blackrock Ranger Station, is remembered at the Rosencrans Cabin Historic District in northwestern Wyoming. This historic district includes the log cabin—known as "Rosie's Cabin" for his nickname—in which Ranger Rosencrans lived and from which he worked for most of his remarkable Forest Service career.

Born in 1875, this son of Austria's chief forester was orphaned at age six and went to live with his older sister in Bohemia. After earning an engineering degree at the University of Vienna and attending Austria's naval academy, he went to sea in an Austrian Navy battleship as a navigator. Yearning for Wyoming since seeing Buffalo Bill Cody's Wild West Show in Vienna as a boy, Rosencrans jumped ship in San Francisco and lit out for the land of his dreams with a seabag-full of technical skills. He became an American citizen, and was hired in 1904 as a forest ranger by the supervisor of what was then the Teton Division of the Yellowstone Forest Reserve and in 1908 became the Teton National Forest.

As the first ranger on the Buffalo Ranger District, "Rosie"—as he was known throughout the Jackson Hole country—and a fellow ranger built the first Blackrock Ranger Station in 1904, and later constructed the ranger station living quarters now called "Rosie's Cabin." From this station, Ranger Rosencrans patrolled an unmapped district that extended about twenty miles in all directions. His district wasn't unmapped for long. Using his technical skills, Rosie drew maps of the country that were praised by the Jackson, Wyoming, *Courier* in 1949 as "models of accuracy and unexcelled draftsmanship." The bane of poachers and even stagecoach robbers who drifted into his district, Rosie was the consummate forest ranger, and everyone knew it. But the May 11, 1907, issue of the Pocatello, Idaho, *Tribune* may have embarrassed him when it gushed:

> A remarkable character is Rudolph Rosencrans . . . (who) has trailed through the cloud-kissed Tetons, ever on the lookout for forest fires and poachers, extinguishing the one and arresting the others. No more efficient ranger ever threw a diamond hitch and a handier man on webs or skis never buckled to a long slope of gleaming snow.

Rosie's public relations skills won local ranchers over to Forest Service methods of grazing national forest range. And the Blackrock Ranger Station cabin he built housed the Buffalo District ranger for decades and Forest Service summer employees for decades more.

But, at age fifty-five, failing eyesight—perhaps a result of drawing all those intricate maps in poor light, or of the glare of the sun on snow, or both—forced Rosie to retire from the Forest Service in 1928. He lived another forty-two years in Jackson, the last fifteen in blindness before he died in 1970 at ninety-four. Retired Ranger Rudolph "Rosie" Rosencrans was laid to rest by

Ranger Rudolph "Rosie" Rosencrans in 1924 (left) and in his later years (right). U.S. Forest Service photos.

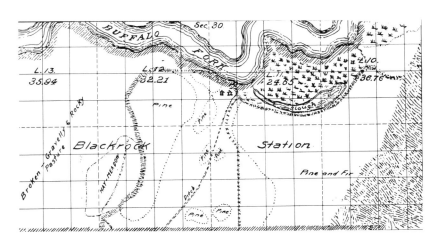

A map of the Blackrock Ranger Station vicinity drawn by Ranger Rudolph "Rosie" Rosencrans.

the Buffalo Fork of the Snake River—near the cabin that bears his nickname and within the Rosencrans Cabin Historic District.

ACCESS: Rosencrans Cabin Historic District is just off U.S. Highway 26-287, about 1/2 mile north of the current Blackrock Ranger Station and about 12 miles east of Moran Junction. It is an easy side trip for visitors to Yellowstone and Grand Teton national parks. Tours of the historic district may be arranged by contacting the Blackrock Ranger Station, P.O. Box 278, Moran, Wyoming 83013; telephone (307) 543-2386.

TONY GROVE RANGER STATION
Wasatch-Cache National Forest, Utah
(1907)

Tony Grove Ranger Station, which began as a log cabin built as a Cache National Forest district ranger's office in 1907, has been in continuous use ever since. The 1907 cabin, now the centerpiece of the Tony Grove Memorial Ranger Station Historic District, and several wood-frame structures built in the late 1930s have been used as a ranger station, forest tree nursery, forestry school instructional facility, and guard station. And now, even as it houses the Forest Service recreation guard for the Tony Grove Canyon Recreation Area, it's being groomed for still another use—interpretation as a Logan Canyon Scenic Byway historic site.

The historic ranger station cabin was built about the time the forest reserves were renamed national forests. A one-story, two-room structure of saw-hewn logs, half-dovetail notched at the corners, it has a log and plank front porch and a moderately-pitched roof covered with dark green cedar shingles. Surrounded by aspen and willow stands broken by meadows, the cabin looks like the ranger station it was—a station from which early rangers armed with the *Use Book* and good sense worked to protect and manage the district's range, timber, and other resources.

During the Great Depression, when massive New Deal public works programs employed thousands to rehabilitate logged and

A forest officer uses the telephone on the porch of Tony Grove Ranger Station, Cache National Forest, Utah, in 1937 (above, U.S. Forest Service photo) and Tony Grove Memorial Ranger Station, Wasatch-Cache National Forest, Utah, today (below).

burned public lands, a tree nursery was established at Tony Grove to produce seedlings to replant national forests in Utah and Idaho. Between 1937 and 1939, the Civilian Conservation Corps constructed several wood-frame buildings and other nursery facilities. They remained to help operate the nursery. By 1939, using seed from several Intermountain Region national forests, one and a half million seedlings were started at the Tony Grove nursery, and by 1942, nursery stock was delivered to fourteen Region 4 forests. Trees grown at the nursery included ponderosa and lodgepole pine and Engelmann and blue spruce. Some nursery operations continued into the 1950s. A few remaining CCC structures—notably, a six-horse barn and a generator shed—and a dense stand of ponderosa pines planted in rows behind the cabin are remnants of the nursery era. The cabin, rehabilitated in the 1970s, has been well-maintained.

Through much of its history, the Tony Grove facility maintained close ties with forestry faculty and students at Utah Agricultural College, now Utah State University, and Utah State students still receive forestry instruction at the site and on adjacent national forest lands.

Today, with development of the Logan Canyon Scenic Byway, historic Tony Grove Ranger Station's role in interpreting the Forest Service's history and mission is expanding.

ACCESS: Historic Tony Grove Ranger Station is 23 miles northeast of Logan on U.S. Highway 89, the Logan Canyon Scenic Byway. Turn left off U.S. Highway 89 at the sign to Tony Grove Canyon Recreation Area, and follow the paved road for 1/4 mile to the public parking area.

INDIANOLA RANGER STATION
Salmon National Forest, Idaho
(1909)

It was an inauspicious beginning. The log cabin Ranger Paul Walter lived in as the first Forest Service ranger at Indianola—a

The 1928 office building (left) and 1909 residence (right) at historic Indianola Ranger Station, Salmon National Forest, Idaho. U.S. Forest Service photo.

post office and stagecoach stop settlement on the Salmon River that became a ranger station—burned to the ground not long after he got there. It got better. A new residence for Ranger Walter was built at Indianola in 1908 and 1909, and the Forest Service purchased the settlement for $1,000 to serve as a ranger station. After sixty-three years as the ranger's residence—and, for the first ten of those years, office as well—at Indianola Ranger Station, the historic structure remains in service as today's Indianola Field Station office.

By 1926, when Ranger Neale Poynor and his wife, Laura, began their seventeen-year stint at Indianola Ranger Station, the residence—an addition having been built in 1912—consisted of a kitchen, office, bedroom, and screened-in porch on the ground floor and three unfinished bedrooms on the second floor. The house itself was of wood-frame construction with clapboard walls on a foundation of logs and stone under the original structure and concrete under the 1912 addition. Water was first piped into the house in 1930.

This historic photograph of Indianola Ranger Station from the rear shows the American flag (left), usually flown when the ranger was at home, and the ranger's vegetable garden. U.S. Forest Service photo.

Between 1933 and 1935, the Civilian Conservation Corps finished the bedrooms, added the front porch, built the north side addition, and excavated a basement. All these additions give the two-story house an irregular shape.

An office for Ranger Poynor, built just north of the residence in 1928, also remains in use. This wood-frame, 25-foot by 35-foot, one-story building is set on a concrete foundation, has walls of shiplap siding, and has a cedar-shingled gable roof. A matching woodshed was built in 1932.

Eight additional buildings were added to Indianola Ranger Station by the CCC beginning in 1933, and all still stand. These structures—which include a bunkhouse, a warehouse, a barn, a large garage-equipment shed, and several other outbuildings—supported the increased workload on the Indianola Ranger District until 1972 when it was incorporated into the North Fork Ranger District. Today, as the Indianola Field Station, historic Indianola Ranger Station—all its buildings painted white with forest green trim—offers a glimpse of Forest Service history even as it serves as a base for summer helicopter crews and other seasonal operations.

ACCESS: To visit historic Indianola Ranger Station, turn west off U.S. Highway 93 at North Fork onto Forest Road 030. Follow paved Forest Road 030 for about 10 1/2 miles down the Salmon River to the station, which is on the north side of the road.

CALIFORNIA BAR RANGER STATION
Salmon National Forest, Idaho
(1909)

California Bar Ranger Station, a log house and barn on a timbered hillside above Napias Creek and the road to Leesburg in the rugged Salmon Mountains of central Idaho, was one of the first permanent ranger stations built on the Salmon National Forest. Thanks to repairs by the Civilian Conservation Corps in the 1930s and Forest Service employees and volunteers in the 1980s, it remains the Salmon's oldest intact ranger station.

The first structure at California Bar Ranger Station was built in 1909—only two years after President Theodore Roosevelt had established the Salmon National Forest—of hand-hewn logs on a site set aside in 1907 as the Leesburg Administrative Site. This building, a one and a half-story, three-room log cabin that doubled as the ranger's residence and office, measures about 25 feet by 30 feet. Its ponderosa pine log walls, round on the outside and hewn flat on the inside, square notched and spiked at the corners, were set on wooden blocks. Its roofs, a gable roof over the main part of the cabin and a shed roof over the back part, were built of sawed lumber and covered by cedar shingles. The second structure, a one-story log barn that measures about 16 feet by 28 feet, was built west of the cabin in 1912 to house Forest Service stock. Set on a fieldstone foundation, the barn has three stalls, each with a grain box and hay manger, and a hayloft.

When first built, California Bar Ranger Station was the center of a large, virtually roadless ranger district. Ranger Gus Schroeder

Historic California Bar Ranger Station, Salmon National Forest, Idaho. U.S. Forest Service photo.

rangered the Leesburg Ranger District from this station until 1915, when he was replaced by Ranger R.L. Dryer who, in turn, was replaced in 1918 by Ranger C.L. McCracken. Its location close to Leesburg, a gold mining town that prospered from the 1860s into the early twentieth century, reflects the close relationship between the Forest Service and mining on national forest lands in those days. Officially renamed California Bar—after a nearby Napias Creek sand bar that probably reminded early placer miners of the California gold rush, so they named it accordingly—in 1936, the station continued as an administrative outpost for the Salmon National Forest into the 1940s.

Maintenance by the CCC in the 1930s—they gave the cabin a concrete foundation, replaced its rotting sill logs, rebuilt its roof, and also re-roofed the barn—kept the station in shape. But, by the 1950s, the station was used only intermittently, and maintenance ended. By the 1980s, the station was in disrepair. As part of a Forest Service program to train volunteers to preserve pre-1940s log structures, Forest Service employees and volunteers made major

Barn at historic California Bar Ranger Station.
U.S. Forest Service photo.

repairs to both the cabin and barn during the 1980s. Both structures are rare examples of early Forest Service log architecture, and the cabin is now used as a back country retreat for hikers.

ACCESS: Twelve miles due west of Salmon, Idaho, historic California Bar Ranger Station is about 3 miles south of the historic gold mining town of Leesburg. That's as the crow flies. On the ground, it's quite a trip to California Bar, which is about 25 road miles from Salmon and about 20 road miles from U.S. Highway 93. The good news is the roads, while not paved, are good all the way. To get there, drive about 5 miles south of Salmon to the junction of U.S. Highway 93 and Williams Creek Road. Turn west onto Williams Creek Road, cross the Shoup Bridge soon, and enter the Salmon National Forest after about 5 miles. Continue on Williams Creek Road, which signs also identify as Forest Road 21, over Williams Creek Summit, which is about 10 miles from both U.S. Highway 93 and California Bar, and continue on Forest Road 21, now called Moccasin Creek Road, to a major new intersection marked by a

sign that reads "ENTERING BEARTRACK GOLD MINE PRO-
JECT." Turn right, or north, onto this new, wide road that goes
all the way to Leesburg. Follow it for a few miles, and see historic
California Bar Ranger Station on the left about 3 miles south of
Leesburg.

POLE CREEK RANGER STATION
Sawtooth National Forest, Idaho
(1909)

Seemingly lost in the blue-green vastness of the Sawtooth Valley, historic Pole Creek Ranger Station—called Pole Creek Guard Station later in its history—is the oldest Forest Service constructed building on the Sawtooth National Forest. Ranger William H. "Bill" Horton, who served at Pole Creek every summer from 1908 to 1929, contributed most of the labor that converted $308.00 in materials and supplies into this 1909 cabin. The site along Pole Creek, named for the lodgepole pines that grow along its course, was officially reserved for ranger station use because of its location. In addition to its proximity to existing travel routes, it had a good water source, grazing for livestock, and a great view.

Working from this station, which also included a horse barn and an equipment shed that have long since disappeared, Ranger Horton did an early-days forest ranger's job. His official diary for the summer of 1915, for example, reflects a wide variety of duties. Some were housekeeping chores, including shoeing horses and packing equipment. Others, such as locating telephone lines, developed the national forest's infrastructure. Still others, posting range boundaries and checking grazing allotments, were resource management jobs. And fire protection involved fire patrols and fire fighting.

A self-guided tour of historic Pole Creek Ranger Station leads along paved, wheelchair-accessible walkways about 300 yards across Pole Creek to the well-preserved ranger station cabin, and a shorter distance past rest rooms to a scenic overlook. Interpretive signs along the way tell the story of the station and its first ranger.

Historic Pole Creek Ranger Station on Pole Creek, Sawtooth National Forest, Idaho.

ACCESS: Historic Pole Creek Ranger Station is located 3 miles east of Idaho Highway 75 on Pole Creek Road, a well-signed secondary gravel road that leaves the highway 24 miles south of Stanley and 35 miles north of Ketchum.

HAYS RANGER STATION
Payette National Forest, Idaho
(1913)

The fires that raged on the Payette National Forest in 1994 almost claimed historic Hays Ranger Station, a log cabin built by the Forest Service in 1913 that served as a ranger station on the old Idaho National Forest until 1918. But firefighters foamed the cabin and saved it from the so-called Chicken Complex Fires that burned tens of thousands of surrounding acres. After the fire had passed and the smoke had cleared, historic Hays Ranger Station still stood in what one Forest Service officer called "a lush green patch in the middle of the burn."

Hays Ranger Station began, not as a ranger station, but as a homestead. Charles B. Hayes, an early homesteader, built a log cabin on level ground above the South Fork of the Salmon River around 1900. He lived there until 1908, when he turned his land rights over to the Idaho National Forest and the Forest Service acquired his cabin. Hays Ranger Station—the station took Hayes's name, but over the years the "e" was lost—served as a Warren Ranger District station from 1908 to 1918. A fire in 1913 burned the log cabin Hayes built, and Forest Service personnel constructed another log cabin on the same spot. That is the cabin that survived the 1994 fires and remains the oldest standing Forest Service building on the Payette National Forest created in 1944 by combining the old Idaho and Weiser national forests.

By 1918, the Forest Service had abandoned Hays Ranger Station and set up another ranger station in Warren, about ten miles to the northwest. From 1918 into the 1950s, a variety of unauthorized occupants and hunters used the old Hays Ranger Station cabin

Historic Hays Ranger Station, Payette National Forest, Idaho. U.S. Forest Service photo.

that, eventually, fell into disrepair and ruin. The cabin was restored in 1988, and is maintained as an historic site.

China Mountain's southeastern slope, where historic Hays Ranger Station still stands, features more history. Charles Hayes wasn't the first to settle there. Before he came, from about 1870 to 1890, Chinese sojourners farmed the land, terracing the slopes and irrigating vegetable gardens and hay fields. Their produce helped feed the population of the Warren mining district. When the mines played out, the Chinese left. Hayes settled on the developed agricultural land abandoned by the Chinese gardeners. Remnants of the Chinese gardens, including a reconstructed replica of farmer Ah Toy's dugout dwelling, also are interpreted at the Hays Ranger Station site.

ACCESS: *Historic Hays Ranger Station is located about 10 miles southeast of Warren, Idaho, on Forest Road 337. To get there, follow Forest Road 340 southeast from Warren for about 4 miles, then bear left onto Forest Road 337 and follow it to the Hays Ranger Station site. Warren is about 52 miles northeast of McCall via Forest Road 21.*

Historic Elk Creek Ranger Station, Bridger-Teton National Forest, Wyoming. U.S. Forest Service photo.

ELK CREEK RANGER STATION
Bridger-Teton National Forest, Wyoming
(1914)

A small log cabin, built in 1914 and known as Elk Creek Ranger Station and later as Elk Creek Guard Station, is typical of Forest Service ranger stations built in the early 1900s. Not much is known of its history, but a visitor to this lonely site on the Hams Fork River gets a feel for the rigors of the early forest ranger's life and work there.

ACCESS: The cabin that was Elk Creek Ranger Station stands along Forest Road 10069, a gravel road that follows the Hams Fork River north from Kemmerer, a southwestern Wyoming town on U.S. Highway 30, and is about 35 miles north of Kemmerer. An interpretive sign identifies the cabin as an "Early Day Ranger Station."

Hewinta Guard Station, Wasatch-Cache National Forest, Utah. U.S. Forest Service photo.

HEWINTA GUARD STATION
Wasatch-Cache National Forest, Utah
(1928)

Hewinta Guard Station, built in 1928 by a tie hack, was a U.S. Forest Service administrative site toward the end of the tie hack era. This cabin, which illustrates the skill of the tie hack woodsman, is a vestige of a national forest use now long gone. Indeed, the real story of Hewinta Guard Station is not at the station itself, but in the Uinta Mountains country. It is the story of the tie hacks.

Who was a tie hack? What was the tie hack era?

A tie hack was a woodsman who, from the late 1800s until about 1940, lived in the forest year round while hand hewing—or hacking—railroad ties from logs. The ties they produced made possible the laying of the railroad tracks that opened up the vast resources of the West.

Most tie hacks on and around the old Wasatch National Forest were Scandinavian woodsmen, but at times local farmers, ranchers, and sheepmen worked hacking ties to make extra money. Ties were

138

hewn year round, hauled out of the forest by horse-drawn skids or sleds to decking areas and, during spring thaw, floated down the Blacks River and other streams or flumes to railroad yards.

Some tie hack woodsmen lived in isolated bachelor camps while others lived in company compounds that included bachelor cabins, bunkhouses, a cookhouse, and a barn. The few married tie hacks lived in cabins with their families. Tie hacks built their own cabins. It took about twenty-two days for one man to build the typical one-room, 10-foot by 12-foot cabin of nine-course log walls topped by a low-gabled, flat-board roof with a dirt insulation barrier. Tie hacks moved their cabins from job to job simply by taking the cabin apart at one site and rebuilding it at another.

About 1940, sawmills and motorized transportation ended the tie hack era. Today, abandoned log cabins, many of which belonged to tie hacks, spot the north slope of the Uinta Mountains. Hewinta Guard Station—that name, by the way, is a combination of the verb "to hew" and the name of the mountain range—was a base for Forest Service administration of the tie hacking industry and, later, other resource uses, on the old Wasatch National Forest.

An old tie hack cabin, built in 1927, was moved to Mountain View Ranger Station—north of Hewinta Guard Station in Mountain View, Wyoming—in 1990 and restored. This move and restoration, made possible by a partnership between the Forest Service and TCI Cablevision of Wyoming, provides a close-up look at the tie hack's life.

ACCESS: Hewinta Guard Station is about 30 miles south of Mountain view, Wyoming, which is just 6 miles south of Interstate Highway 80 in southwestern Wyoming on Wyoming Highway 414. To reach Hewinta Guard Station, follow Wyoming Highway 410 southwesterly from Mountain View for 10 miles to Robertson, then follow Forest Road 075 into the Wasatch-Cache National Forest and across the border into Utah for 10 miles to the junction with Forest Road 058. Turn right and follow Forest Road 058 for 3 miles to Forest Road 206, turn left and proceed another mile to Hewinta Guard Station. Back in Mountain View, at Mountain View Ranger Station, see the tie hack cabin display and learn more about tie hacks on the national forest.

The 1933 office building at Markleeville Guard Station, Toiyabe National Forest, California. Photo by Nancy Thornburg.

MARKLEEVILLE RANGER STATION
Toiyabe National Forest, California
(1933)

As small and out of the way as Markleeville, the diminutive seat of Alpine County—California's least populous—is, its three historic ranger stations have played a part in the history of three national forests and two Forest Service regions. Although the first of those ranger stations, Hot Springs Ranger Station west of town on Hot Springs Creek near what is now Grover Hot Springs State Park, no longer exists, the second, now a private residence on Montgomery Street, and the third, now Markleeville Guard Station, remain Markleeville landmarks.

Hot Springs Ranger Station, built about 1905 when the Markleeville country was part of the Stanislaus National Forest, became a Mono National Forest ranger station when that forest was established in 1908 from parts of the Stanislaus, Tahoe, Sierra, and Inyo national forests. Other rangers on the Mono National Forest were posted at Bridgeport and Lee Vining, California, and Sweetwater, Nevada.

The old Hot Springs Ranger Station, east of Markleeville, as it appeared about 1918 (above), and the first Markleeville Ranger Station, on Montgomery Street, as it appears today (below). William J. Clark collection and Nancy Thornburg photos courtesy of Alpine County Museum.

William J. "Bill" Clark, ranger of the Alpine Ranger District from 1915 to 1935, began his career at Hot Springs Ranger Station under William M. Maule, a Cornell University forester who became Mono National Forest supervisor in 1909 and held that post until he retired in 1938. At some point, Ranger Clark moved his headquarters to the first "in-town" Markleeville Ranger Station, the Montgomery Street house. Ranger and Mrs. Clark and their family lived in the Markleeville house during the summer and fall, and in the winter and spring they lived in Minden, Nevada, where Ranger Clark worked out of Forest Supervisor Maule's office. In July 1933, Ranger Clark negotiated the U.S. Government's purchase from Alpine County of a town lot in Markleeville for a new ranger station. Uncle Sam paid ten dollars for the lot.

Construction of the second in-town Markleeville Ranger Station began at the end of August 1933. Forest Supervisor Maule and California Region ranger station architect Norman Blanchard traveled to Markleeville on the 29th to inspect the site and building materials that had been delivered. Mr. Maule picked up the deed for the land from the Alpine County clerk that same day, and arranged for the arriving Civilian Conservation Corps crew to begin work the next day. The new Markleeville Ranger Station—an office, a three-bedroom ranger's residence and a smaller residence, a garage, a warehouse, a blacksmith shop, and outbuildings—was Ranger Clark's headquarters until he retired in 1935. Ranger Bill Hays replaced him.

Two years later, a December flood swept through Markleeville—nestled in a tiny, pine-sloped basin bisected by Markleeville Creek—and Markleeville Ranger Station. That 1937 flood was the first of several. Others occurred in 1950, 1953, and 1963. Also, during those years, some of the station's now-historic buildings were relocated on the site, and other buildings were built.

Markleeville Ranger Station became Markleeville Guard Station in 1939 when the Alpine Ranger District's office was moved to Minden, just after the Mono National Forest's supervisor's office was moved to Reno. Then, in 1945, the Mono National Forest was split between the Inyo National Forest and the Toiyabe National Forest, and this ranger station built to California Region standard

The ranger's residence at Markleeville Ranger Station during the December 1937 flood. Photo from Ray Koenig collection courtesy of Alpine County Museum.

plans—along with the old Bridgeport, California, ranger station—wound up in the Intermountain Region.

The station's history got even more complicated in 1973 when the Toiyabe National Forest's Minden-based Alpine Ranger District—along with Markleeville Guard Station—was absorbed by the forest's Carson City-based Carson Ranger District. Today, historic Markleeville Guard Station—a link to the early days—is an interagency station operated by the Toiyabe National Forest for fire prevention, fire suppression, visitor information, and other forest management purposes.

ACCESS: Easily identified by its Markleeville Guard Station sign, historic Markleeville Ranger Station is located on California Highways 4 and 89 in Markleeville, California, about 25 miles south of Minden, Nevada. Visitors are welcome at the office, usually open from about June 1 through October on Thursdays through Mondays from 8:00 a.m. to 4:30 p.m. Fire crews at the station are happy to show visitors their wildland fire engines and explain wildland fire

143

prevention and firefighting. The nearby Alpine County Museum, open from Memorial Day weekend through October every day except Tuesdays from 12:00 noon to 5:00 p.m. or by appointment— call (916) 694-2317—also welcomes visitors.

VALLEY CREEK RANGER STATION
Sawtooth National Forest, Idaho
(1933)

Valley Creek Ranger Station, the first Forest Service ranger station in the Stanley Basin of central Idaho, was established by the new Sawtooth National Forest in 1908 near the confluence of Valley Creek and the Salmon River.

The original ranger station building, a three-room log house completed in 1909 while either Edgar P. Huffman or Wallin Job was ranger—official records and old-timer recollections differ on this point—and transferred to the Challis National Forest in 1913, was headquarters for several Stanley Basin Ranger District rangers until 1933. In that year, at about the middle of Ranger Merle Markle's nine and one-half years there, a new ranger station main building begun the previous year was completed. The old building was sold to Dave Williams, a Sawtooth Valley rancher, who moved it into Upper Stanley for his family to live in during the school year.

Ranger Markle cut and hauled the logs for the new Valley Creek Ranger Station building, and his wife, Kathleen, helped with the peeling. A crew of men helped with the actual construction. The new building, completed at a cost of $3,000, had nine rooms and a bath. Ranger Markle, who previously had built a telephone office-garage in 1929 and a machine shop in 1932, rounded out the ranger station compound by supervising construction of a barn in 1933, a woodshed-cellar in 1934 and 1935, and an oil house in 1936. All but the machine shop and oil house remain. Along with the main building, the garage and telephone office are built of logs and the barn has log siding.

Eight district rangers followed Ranger Markle at Valley Creek Ranger Station. Tom Kovalicky, later Nez Perce National Forest

Historic Valley Creek Ranger Station, Sawtooth National Forest, Idaho, is now the Stanley Museum.

supervisor, was the last. Stanley Ranger Station, a new headquarters for the Stanley Basin Ranger District, was completed on U.S. Highway 93 two and one-half miles south of Stanley in 1970. In 1972, the district and historic Valley Creek Ranger Station were transferred from the Challis National Forest to the newly-created Sawtooth National Recreation Area of the Sawtooth National Forest.

Today, the Sawtooth Interpretive and Historical Association operates Stanley Museum, a local pioneer museum featuring the settling of the Sawtooth Valley and the Stanley Basin, in the historic Valley Creek Ranger Station main building.

ACCESS: Historic Valley Creek Ranger Station is about 1/2 mile north of Stanley, Idaho, on the west side of Idaho Highway 75. Stanley is reached by following Idaho Highway 21 from Boise, U.S. Highway 93 and Idaho Highway 75 north from Twin Falls, and U.S. Highway 93 and Idaho Highway 75 south from Missoula, Montana. Stanley Museum, in the old Valley Creek Ranger Station, is open from late May through Labor Day.

Warren Guard Station, Payette National Forest, Idaho.
U.S. Forest Service photo.

WARREN GUARD STATION
Payette National Forest, Idaho
(1934)

In continuous use for more than sixty years, the Warren Guard Station office building, erected in 1934, was one of the first Civilian Conservation Corps projects in Idaho. Still in service, this building is a well preserved example of the standard designs the Forest Service developed to meet its growing need for administrative sites in the 1930s and a testimony to the excellent work of the CCC. Warren Guard Station itself dates from 1918 when the Forest Service moved from the old Hays Ranger Station (pages 135 to 136) to Warren.

Within the limits of the historic gold mining town of Warren, Warren Guard Station fronts on the old Warren Wagon Road, now known as Forest Road 21. Although a wood-frame structure on a concrete foundation, this Rocky Mountain Cabin style office building sports Shevlin log cabin siding developed by the Shevlin-Hixon

Lumber Company of Bend, Oregon, to simulate log construction. Classical touches include a temple front with columns. An antique water cannon, once used in Warren area placer mining operations, is on display.

The office building is surrounded by other Warren Guard Station buildings. Three of these—a residence, a bunkhouse, and a barn—add to the station's historic significance. The barn, a 1909 peeled log structure with dovetail notching, is especially interesting.

ACCESS: Warren Guard Station is located on Forest Road 21 in the town of Warren, and the office is open during normal business hours from June through early September. Warren is about 52 miles northeast of McCall, Idaho, via Forest Road 21.

SQUIRREL MEADOWS GUARD STATION
Targhee National Forest, Wyoming
(1934)

Squirrel Meadows Guard Station, constructed by the Forest Service in 1934 on a northwestern Wyoming administrative site set aside in 1907, is typical of Intermountain Region field cabins built to standard plans in the 1930s. After decades of use by Targhee National Forest rangers and guards, this historic station is one of about forty rustic guard stations and fire lookouts in the region now rented to national forest visitors.

The cabin and outhouse that are Squirrel Meadows Guard Station replaced an earlier facility of which nothing—not even a description—remains. It may have been a cabin that occupied the site at the time it was withdrawn, or a cabin built by the Forest Service at some later date. At any rate, the 1934 cabin is a two-room, 32-foot by 18-foot log structure that has a six-foot-wide porch on its east end. The native lodgepole pine walls are single saddle notched and chinked with split poles. Originally finished with a linseed-base stain on the logs walls and gray trim on the window frames, the cabin is now stained brown with green trim.

Squirrel Meadows Guard Station, Targhee National Forest,
Wyoming. U.S. Forest Service photo.

A light green metal roof covers the original wood shingles. The foundation is concrete.

A well, dug immediately east of the cabin in 1981 and capped with a hand pump, provides water. Eighty feet south of and across the road from the cabin is the outhouse, also built to a standard 1930s plan. Since such necessaries were moved as their pits were filled, this outhouse probably is not in its original location. An intrusive white propane tank that fuels lighting and appliances barely mars the historic aura.

The lodgepole pines that once surrounded Squirrel Meadows Guard Station were cut to combat a mountain pine bark beetle infestation, and the station is now surrounded by a grassy opening and faces a large camas meadow called Squirrel Meadows. Squirrel Creek, a perennial stream bordered by willows, flows immediately east of the station. The cabin's porch, which faces east, affords an excellent view of the meadows, the creek, and the northern Teton Range about six miles distant. The east entry was probably the cabin's original main entry, but the west entry, which opens directly toward the road to the station, is now the principal entry.

ACCESS: Squirrel Meadows Guard Station is about 22 miles due east of Ashton, Idaho, and about 2 miles inside Wyoming on Forest Road 261, the Reclamation Road/Ashton-Flag Road. To get there, turn east off U.S. Highway 20 in Ashton onto Idaho Highway 47, then turn south onto Idaho Highway 32 about 1/4 mile east of town. Drive 1 mile south on Idaho Highway 32, then turn east onto Forest Road 261 and follow that road for about 22 miles to Squirrel Meadows Guard Station. Contact Ashton Ranger District, 46 U.S. Highway 20, Ashton, Idaho 83420, telephone (208) 652-7442, for Squirrel Meadows Guard Station rental information and current road conditions.

Thirteen historic U.S. Forest Service ranger stations of the Pacific Southwest Region (Region 5) which comprises most of California.

Chapter Five
PACIFIC SOUTHWEST REGION

California's eighteen national forests comprise the Pacific Southwest Region, also called Region 5 and once named the California Region. These forests, like the state itself, are a land of contrasts. Wrapped around the Great Central Valley, these national forests—in the higher elevations of the Coast Ranges on the west, the Klamath Mountains, Cascade Range, and Modoc Plateau to the north, the Sierra Nevada and Basin Ranges on the east, and the Transverse and Peninsular ranges to the south—encompass an environmental diversity that affords an abundance of natural resources. Within the more than eighteen million acres of California's national forests are the chaparral, oak and coniferous woodlands, and coniferous forests—as varied as the coastal redwood rain forest, the Douglas-fir stands of the northwest, and the yellow pine belt of the Sierra—that provide vital water as well as timber, grazing, recreation, abundant natural beauty, and other benefits enjoyed by citizens of and visitors to the nation's most populous state. These forests are managed by a regional forester in San Francisco, seventeen forest supervisors, and about seventy-five district rangers.

Forest rangers began to protect these forests in the late 1890s, and built California's first ranger station—West Fork Ranger Station on the San Gabriel Forest Reserve—in 1900. Several of Region 5's relatively few remaining pre-Depression ranger stations, and a variety of Depression-era ranger stations, are profiled in this chapter. Despite efforts as early as 1917 to standardize its ranger stations, the Pacific Southwest Region's historic ranger stations are as diverse as its forests.

Rangers Louie Newcomb, Phil Begue, Willard Sevier, and Jack Baldwin at West Fork Ranger Station, San Gabriel Forest Reserve, California, shortly after completion in 1900. Photo courtesy of Big Santa Anita Historical Society.

WEST FORK RANGER STATION
Angeles National Forest, California
(1900)

The rangers who patrolled California's first forest reserve, set aside as the San Gabriel Timberland Reserve by President Benjamin Harrison in 1892, soon needed California's first federal ranger station, West Fork Ranger Station built on the West Fork of the San Gabriel River in 1900. Today, surrounded by a protective wood rail fence at the Chilao Visitor Center less than ten miles from its original site, the historic West Fork Ranger Station cabin gives Angeles National Forest visitors a glimpse of turn-of-the-century ranger life.

These early U.S. Department of the Interior forest rangers, charged with protecting the forest reserves before establishment of the Forest Service and the National Forest System, found their

The abandoned West Fork Ranger Station, Angeles National Forest, before it was moved to Chilao Visitor Center. Photo courtesy of Big Santa Anita Historical Society.

effectiveness limited by beginning their patrols from communities outside the reserve. The solution was to build a base of operations— a continuously staffed ranger station from which they could respond to fires and other emergencies in a more timely way—inside the reserve. Ranger Louie Newcomb supervised construction of West Fork Ranger Station for seventy dollars in federal funds in 1900, and stayed on as ranger in charge for three years. During those years, in addition to fighting fires, building trails, and helping lost hikers and fishermen, he helped build the San Gabriel Forest Reserve's ranger station at Pine Flat—renamed Charlton Flat in 1925—in 1902 and built a ranger station at Sturdevant's Camp in 1903.

Ranger Newcomb's cabin at West Fork Ranger Station was a well-built, one-room structure of about fifteen by twenty feet, similar to others he had built in the area. Alder logs for the walls were cut from the creek bottom near the cabin's original location, incense cedar for the roof came from the surrounding hillsides. A

Historic West Fork Ranger Station at Chilao Visitor Center, Angeles National Forest, California. U.S. Forest Service photo.

stove provided heat. Outside were a corral, fencing, and a hitching post for the rangers' saddle and pack stock. For over twenty-five years, this cabin was both office and home for General Land Office and, after 1905, Forest Service rangers.

Closed after the Devil's Fire of 1924, the station cabin was reoccupied in 1936 after a new ranger station was built next to it in 1933. A granite foundation, cement floor, and second roof seem to have been added at that time. Again it played a role in protecting the Angeles National Forest. Eventually, the cabin was used as a storage building at the newer ranger station. But when that station became obsolete and was dismantled in 1977, the old West Fork Ranger Station cabin was abandoned. Threatened by insects, dry rot, and vandalism, it was rescued and reconstructed behind the Chilao Visitor Center in 1983. Preserved as the oldest standing ranger station on the Angeles National Forest and, almost certainly, on all of California's national forests, it is outfitted with period artifacts to show what life as a ranger during the early years was like.

ACCESS: To visit historic West Fork Ranger Station at Chilao Visitor Center, turn off Interstate 210 at La Canada onto California Highway 2, the Angeles Crest Highway, which also is designated the Angeles Crest Scenic Byway, and follow that highway for about 27 miles into the Angeles National Forest to the Charlton-Chilao Recreation Area. Follow the signs to the second marked Chilao entrance, and follow the Chilao Road for about 1/2 mile to the Chilao Visitor Center. The historic West Fork Ranger Station cabin is located directly behind the visitor center, which is open Fridays through Mondays from 9:00 a.m. to 5:00 p.m.

PYRAMID RANGER STATION
Eldorado National Forest, California
(1910)

Pyramid Ranger Station, built along the American River between Placerville and Lake Tahoe between 1910 and 1911 and used as a guard station into the 1970s, is the oldest standing station on the Eldorado National Forest. But, since the fall of 1993, it's been standing elsewhere—at the Owens Camp Fire Station about four and a half miles southwest of its original site. There, the Forest Service fire crew is to spend the next few years rehabilitating and refurnishing the building to serve as an office for the engine captain and an interpretive center for visitors.

As the Eldorado National Forest sees it, the Owens Camp crew "will become the new 'forest guards,' and will serve the public as 'rangers' of the past" at the relocated and restored Pyramid Ranger Station building, "providing information, issuing permits, enhancing and protecting the resources, and fighting fires" as they "convey a sense of the early days of rangering" on the national forests.

This Pyramid Ranger Station tradition the Owens Camp crew is to carry on began with a 1908 proposal by E.L. Scott, a Tahoe National Forest assistant ranger, to build a ranger station at the American River site "because no other tract in this vicinity is suitable. . . ." Within a month, the paper trail shows, the district forester in San Francisco had approved and requested that the site

Early-day rangers at Pyramid Ranger Station, Eldorado National Forest, California. U.S. Forest Service photo.

"be withdrawn from settlement, entry, and all other forms of appropriation under the public land laws." Once the Commissioner of the General Land Office granted that request, which he did, the Forest Service was free to build. By the time the single-story, wood-frame "shotgun" building measuring about 38 feet by 16 feet was completed—apparently by forest guards, with all lumber and materials purchased locally and transferred to the site by wagon—in 1911, it belonged to the Eldorado National Forest, established in 1910 from parts of the Tahoe and Stanislaus national forests. At completion, Ranger R.C.M. Berriman valued Pyramid Ranger Station improvements, which also included a barn and fencing, at $900.

Eldorado National Forest rangers and fire guards carried out their duties from this summer station—soon hard along U.S. Highway 50, the main route between San Francisco and Lake Tahoe—for at least the next sixty years. Except for minor improvements and paint scheme changes, the combination office-residence changed little during those decades. In the mid-1930s, the Civilian

The historic Pyramid Ranger Station building was placed on a new foundation at Owen Camp Fire Station, Eldorado National Forest, in 1994 after being moved from its original location in 1993.

Conservation Corps added a bit to the rear of the building and placed it on a concrete foundation. And, also in the mid-1930s, the original barn, by then across U.S. Highway 50 from the rest of the station, was torn down and replaced with the barn that remained on the site after the main building was moved to Owens Camp in 1993.

ACCESS: To visit the historic Pyramid Ranger Station building at Owens Camp Fire Station, turn south off U.S. Highway 50 onto Silver Fork Road at Kyburz, about 30 miles east of Placerville. Follow paved Silver Fork Road for about 4 1/2 miles, turn left onto and follow Bark Shanty Road for about 1/4 mile, then turn left onto and follow Long Canyon Road for about 1/2 mile to Owens Camp. The office, in the historic Pyramid Ranger Station building, is open intermittently from June through mid-October when the fire crew is at the station; when the office is not open, the building may be viewed from the outside.

El Prado Cabin, the former Laguna Ranger Station, Cleveland National Forest, California. Photo by Irene Wiswell courtesy of Laguna Mountain Volunteer Association.

LAGUNA RANGER STATION
Cleveland National Forest, California
(1912)

The log cabin that was Laguna Ranger Station, the "first permanent ranger station in the Laguna Mountains" according to the interpretive sign and now called El Prado Cabin, still stands on the shady slope overlooking El Prado Meadow where Ranger Carl Brenner, his assistant, and some local men built it in a hurry but built it well in 1912.

This old ranger station, a log cabin roughly twelve feet square, seems to have been built within months of a March 1912 tour of the Cleveland National Forest by Assistant District Forester—today his title would be Assistant Regional Forester—Louis A. Barrett. "A one-room log cabin is needed at Laguna Station," he observed, "there being no stopping place in that region." So Ranger Brenner and his crew built the recommended cabin—a dirt-floored, mud-chinked, saddle-notched log structure set on a rock foundation and topped by a pyramidal, split-shake roof—probably intended for no more than seasonal use. A 1914 report described Laguna Ranger Station as "improved with a log cabin and pasture fence and has a good spring."

But more than seasonal use was in Laguna Ranger Station's future. Several years after the cabin was built, the Forest Service opened tracts around El Prado Meadow for summer cabin construction. This was the first step in development of the Laguna Mountain area, a favorite summer resort for San Diegans as well as for Imperial Valley residents seeking to escape the desert heat, for public recreation. As summer recreation increased and Forest Service work expanded, Laguna Ranger Station was in year-round demand. By the 1920s, a more permanent structure was needed. But, instead of replacing it with a new building, the 1912 cabin was improved. A wooden floor and a stone fireplace and chimney were added for year-round comfort. The rock foundation was replaced with concrete. The cabin's corner notches were replaced by upright log corner posts, and the walls were rechinked with concrete. Stabilized and improved, the cabin served as a ranger station through the 1930s.

By 1925, to support increased visitor use of Laguna Recreation Area—which was officially designated in 1926, a log campground registration booth had been erected in front of the cabin.

These early efforts to retain El Prado Cabin for ranger station use helped preserve one of California's few remaining early Forest Service log cabins. As early as 1967, it was maintained as what then-Cleveland National Forest Supervisor Stanley Stevenson termed a "feature attraction for recreation sightseers." Extensive restoration of the cabin in 1989 by members of the Laguna Mountain Volunteer Association under Forest Service guidance ensured El Prado Cabin will remain nestled in the black oaks and Jeffrey pines at the edge of its meadow for years to come.

Hundreds have gathered at El Prado Cabin every Labor Day weekend since 1986 for an annual Living History Weekend sponsored by the Forest Service and the Laguna Mountain Volunteer Association. During this event, visitors are welcomed to the cabin

An "early-day ranger" spins yarns of historic Laguna Ranger Station at El Prado Cabin during the annual Living History Weekend. Photo courtesy of Laguna Mountain Volunteer Association.

by an early-day "ranger" who spins yarns of historic Laguna Ranger Station and the Laguna Mountain country in the old days.

ACCESS: El Prado Cabin is just off the scenic Sunrise Highway, San Diego County Road S-1, and within El Prado Campground in the Laguna Recreation Area of the Cleveland National Forest. To reach El Prado Cabin from San Diego, travel east on Interstate Highway 8 for about 43 miles—about 13 miles beyond Alpine, the closest town that appears on most maps—and exit at the Sunrise Highway. Turn left onto and travel north on the Sunrise Highway for about 14 miles, passing through the community of Mt. Laguna, to the "Laguna-El Prado Campground" sign. Turn left and enter the campground, where there is no fee for visitors who do not plan to camp. There is a campground map on the message board. Drive on, keeping to the right through each intersection, and enter El Prado Campground. El Prado Cabin is next to the road. Park in one of the nearby designated parking areas, not on the road. During the winter, when El Prado Campground is gated, park off the road and walk along the campground road for about 1/4 mile to El Prado Cabin. During the summer months, the Laguna Visitor Information Office, on the Sunrise Highway about 1/4 mile north of the community of Mt. Laguna, provides area information.

ROBINSON FLAT GUARD STATION
Tahoe National Forest, California
(1913)

Robinson Flat Guard Station, built between 1913 and 1916, is among the oldest remaining Forest Service guard stations in California. As the easternmost station on the Foresthill Divide, it was an important base for patrol of and fire control in the upper reaches of the Tahoe National Forest for decades. After roads reached into the area in the 1920s, several popular campgrounds were built near the guard station.

The guard station itself—originally called Robertson Flat before an as-yet-unexplained name change—consisted of a 14-foot by 16-foot wood-frame residence built in 1913, which remains, and a 16-foot by 24-foot wood-frame barn, which was replaced between 1927 and 1930 by a tack and storage shed. A fuel storage shed was added by the Civilian Conservation Corps in 1934.

The appealing residence, both its outside walls and its pyramidal roof covered by wood shingles, housed not only forest guards but also the lookout stationed at nearby Duncan Peak Lookout in the early years. Still in pretty good shape, the one-room structure retains its original tongue-and-groove wood floor, ceiling, and walls as well as its original kitchen cabinets. The front porch, covered by an extended hip roof, is a delightful place to sit and reflect on what it must have been like to serve as the Robinson Flat guard eighty years ago.

Robinson Flat's modern claim to fame is as one of the major checkpoints for the world-famous Western States 100-Mile Run and the Tevis Cup 100-Mile One Day Ride, a footrace and a horse race held each summer along the Western States Trail between Auburn

The barn (left) and residence (right) at Robinson Flat Guard Station, Tahoe National Forest, California, in the old days. U.S. Forest Service photo.

and Squaw Valley. During these events, held in June and August, this normally peaceful meadow teems with activity as crews support their contestants. A hand pump in the meadow was installed to provide cold, pure mountain water for everyone during these events and throughout the summer months.

ACCESS: To visit historic Robinson Flat Guard Station, exit Interstate Highway 80 onto Foresthill Divide Road about 2 miles north of Auburn. Follow this paved road along the Foresthill Divide through the town of Foresthill, 15 1/2 miles from Auburn, and for another 25 miles to the end of the pavement. Then follow a dirt road, Forest Road 43, for about 110 yards past Robinson Flat Campground to the historic guard station.

Historic North Bloomfield Ranger Station,
Tahoe National Forest, California.

NORTH BLOOMFIELD RANGER STATION
Tahoe National Forest, California
(1916)

Just across Humbug Creek from the historic Gold Rush town of North Bloomfield—now the centerpiece of Malakoff Diggins State Historic Park—old North Bloomfield Ranger Station spans the history of the Tahoe National Forest from the early years of the Forest Service through World War II. Even the color schemes of this ranger station's restored buildings, put up over the course of two decades, are faithful to their times of construction.

North Bloomfield Ranger Station was built on land purchased by the Forest Service in 1909, and was headquarters of the old North Bloomfield Ranger District—now part of the Nevada City Ranger District—until 1946. Although the Malakoff Diggins had about played out by the 1880s, North Bloomfield remained a fairly active ranching and timber town when the Forest Service acquired the ranger station site. And the town was a good place for a ranger station. On the ridge separating the drainages of the Middle and

Passport in Time volunteers helped the Forest Service restore historic North Bloomfield Ranger Station (above) and celebrated their contribution on the cookhouse porch (below). U.S. Forest Service photos.

South Yuba Rivers, it afforded good access for the ranger to patrol his district by horse and, later, by car.

The ranger residence-office building combination—now called the cookhouse—and the barn, built during the Tahoe National Forest's early days, recall the time before the California Region imposed standard building plans. Both are representative of structures built during the early twentieth century in the Sierra Nevada foothills where winter snows can be heavy. The 1916 barn, a 47-foot by 27-foot wood-frame structure with board-and-batten siding and three-foot sugar pine shingles on its steep-pitched roof, retains the French-battleship gray color that prevailed in the very early years. The buff-colored cookhouse, a wood-frame structure with wood-shingle siding built in 1931, is one of the last examples of ranger station buildings erected before the California Region adopted Blanchard and Maher's standard designs.

The tobacco-brown ranger residence and garage, constructed in 1934 by the Civilian Conservation Corps, are excellent examples of the region's Depression-era standard designs. The landscaping includes three giant sequoias planted in the island of the compound's oval drive as well as fruit trees and a cork tree.

Restored and preserved by the Forest Service with the help of Passport in Time volunteers, historic North Bloomfield Ranger Station is one of California's few remaining ranger stations that reflect both the early days and the Depression era of Forest Service history.

ACCESS: Historic North Bloomfield Ranger Station is about 10 miles as the crow flies northeast of and about 25 driving miles from Nevada City. To get there, follow California Highway 49 East northwesterly for 11 miles, then turn right onto Tyler Foote Crossing Road at the Malakoff Diggins State Historic Park sign. Follow this paved road for 8 miles to North Columbia, then turn right onto Lake City Road and follow it for 4 miles to Malakoff Diggins State Historic Park and North Bloomfield. In North Bloomfield, turn right at the drug store and follow the paved single-lane road downhill, across the Humbug Creek bridge, and turn right. The historic ranger station is about 100 yards down a gravel road, and parking spaces are clearly marked.

Oak and digger pines framed the newly-completed Sierra National Forest headquarters building in North Fork, California, in 1935. In 1992, as part of North Fork Ranger Station, this building was destroyed by fire. U.S. Forest Service photo.

NORTH FORK RANGER STATION
Sierra National Forest, California
(1919)

Although the historic Sierra National Forest supervisor's office building, centerpiece of California's largest Forest Service compound for almost sixty years, was destroyed by fire in 1992, the compound that became North Fork Ranger Station after the supervisor's office moved to Fresno still harbors many historic structures.

The original Sierra Forest Reserve headquarters may have been old abandoned ranch buildings. North Fork's first Forest Service buildings were built in 1906, the year before forest reserves became national forests. In 1910, the Sierra National Forest's headquarters was moved out of town to a new site surrounding a gently sloping knoll and ridge above Willow Creek northeast of North Fork.

A new headquarters building, constructed atop the knoll in 1910 and 1911 by rangers and other Forest Service personnel, had three rooms. A front room housed the stenographer's desk, switchboard, library case, and drafting board. A supervisor's office and a room used for storage and as a photographic dark room completed the scene. Heat was provided by a furnace under the structure. Water came from a small reservoir via a half-mile pipeline. By 1912, according to the *Sierra Ranger* (the forest's newsletter), the forest supervisor and his deputy were working evenings and Sundays on a "tennis court." In later years, when this building was abandoned, it was converted into the "Pine Tree Club," a community hall and meeting room. This earliest building, unfortunately, no longer exists.

The many historic buildings that remain at North Fork Ranger Station were constructed between 1919 and 1941. Among these are five pre-Depression era residences built between 1919 and 1925 in the Craftsman-Bungalow style popular between 1900 and 1920. A main feature of these structures is their hipped or pyramid roofs. Most of the North Fork Ranger Station compound buildings are standardized designs developed by Blanchard and Maher and constructed beginning in 1933 by men of the Civilian Conservation Corps camp a stone's throw from the compound. These included the main supervisor's office building, the one that burned in 1992, and the supervisor's residence, now used for office space, both built in 1934 and 1935. Three structures built to the same plan as the burned supervisor's office building remain in California today: the former Trinity National Forest headquarters, now Weaverville Ranger Station (pages 175 to 178); the former Shasta National Forest headquarters, now the Mt. Shasta Ranger Station in Mt. Shasta City; and the Plumas National Forest headquarters in Quincy, a much modified version. The old supervisor's residence was built atop the knoll, a location symbolic of its significance.

A total of forty-one buildings were built at the North Fork compound during the Depression era: six in 1933, seven in 1934, eight in 1935, sixteen in 1936, and four in 1939. At least ten of these have been demolished. Also, as these buildings were being built, many of the pre-Depression era buildings were remodeled to reflect the 1930s styles. A comprehensive landscape plan by L.

Glenn Hall, a member of Blanchard and Maher's firm, completed the project. Hundreds of landscape features—roads, parking areas, rock retaining walls and culverts, drainage and septic systems, and native and exotic plants—were built and planted. In 1941, at the close of the CCC program, 2,500 shrubs and 85,760 square feet of lawns were planted, and sprinklers were installed. By the time the United States entered World War II, the North Fork headquarters had developed into one of the largest Forest Service compounds in California. In addition to about fifty buildings, it had a swimming pool and tennis court. This headquarters had become a major part of the local economy, employing many local residents during all or part of the year.

Today, the North Fork Ranger Station compound is the primary office headquarters and work center for the Minarets Ranger District. Its historic buildings and landscape features reflect the first three decades of Forest Service architecture and evoke a strong sense of the past.

ACCESS: North Fork Ranger Station is located in northeastern North Fork, about 35 miles north of Fresno.

PATTERSON RANGER STATION
Modoc National Forest, California
(1921)

During the summers of 1920 and 1921, Ranger Ben L. Johnson of the Modoc National Forest's old South Warner Ranger District built a one-dwelling ranger station in the southern Warner Mountains to support his district's fire suppression and resource management work. The station overlooked Patterson Meadow, named for the Patterson family that settled in the area in 1905 and operated a nearby sawmill that supplied lumber to the eastern Modoc County towns of Cedarville and Eagleville, and was named Patterson Ranger Station. It is now called Patterson Guard Station.

*Historic Patterson Ranger Station,
Modoc National Forest, California.*

After a fire guard named Everett Wilson manned Patterson Ranger Station during its first May to October fire and field season in 1921, the station went unmanned for the next dozen years for want of funds. During those years, when the 216,434-acre district was funded for only a district ranger and one fire guard, the station saw only occasional use. Since 1934, when the Civilian Conservation Corps added a garage, Patterson Guard Station has housed summer fire crews.

Although no monumental Forest Service history seems to have been made at Patterson Guard Station, the station itself is a monument to the old California Region's efforts to standardize its administrative structures. The single-story, wood-frame dwelling Ranger Johnson built on a cement foundation was based on District Forester Coert DuBois's 1917 standard plans. It was the Modoc National Forest's first structure based on those plans, and is the Forest's only surviving pre-1930s building. Maintenance work in 1989 preserved the dwelling. The 1934 garage is based on the region's Depression-era Blanchard and Maher designs. Both buildings and their ponderosa pine forest setting appear about as they did in the 1930s, and the station remains in occasional use by Forest Service fire crews.

ACCESS: Visitors are welcome to view historic Patterson Ranger Station, which occasionally houses Forest Service summer crews, from the outside only. The station is 24 road miles east of Likely, a small town 19 miles south of Alturas on U.S. Highway 395. At Likely, turn east off U.S. Highway 395 onto paved Forest Road 64, the Jess Valley Road. Drive about 9 1/2 miles to the junction with Forest Road 5, bear right at the junction and continue on Forest Road 64 past the Blue Lake turnoff—where the pavement ends—and follow the Patterson Guard Station signs to the reddish-brown station buildings on the left.

INDIANS GUARD STATION
Los Padres National Forest, California
(1929)

Indians Guard Station, built in 1929 in the area of the old Santa Barbara National Forest known as "The Indians," provided a home for forest rangers who patrolled the land and fought fires. During the Great Depression of the 1930s, Civilian Conservation Corps units stationed at Indians constructed the road along Arroyo Seco River. Today, restored Indians Guard Station allows Los Padres National Forest visitors to experience the Forest Service of the 1930s and 1940s. Exhibits feature early fire management and administration. The station houses a volunteer campground host.

ACCESS: Indians Guard Station is 40 road miles west of King City, California, in the Santa Lucia Range. To get there, exit U.S. Highway 101 onto the Jolon Road just north of King City. Drive about 20 miles to Jolon, enter Fort Hunter-Liggett Military Reservation, and proceed to Del Venturi Road, a paved two-lane road. The historic station is located at the end of Del Venturi Road, near

Historic Indians Guard Station, Los Padres National Forest, California. U.S. Forest Service photo.

Santa Lucia Memorial Park Campground. Drinking water is not available, and should be brought by visitors.

GASQUET RANGER STATION
Six Rivers National Forest, California
(1933)

The historic part of the Gasquet Ranger Station compound, the Gasquet Ranger Station Historic District on three acres of a river terrace between U.S. Highway 199 and the Middle Fork of the Smith River, was built by the Civilian Conservation Corps for the Forest Service between 1933 and 1939. Seven CCC-constructed buildings and the low rock wall that surrounds much of the historic compound remain in use and attest to the quality of CCC craftsmanship . . . and to the taste of Ranger Adolph Nielson's wife.

In April 1933, when Adolph Nielson arrived from the Forest Service regional office in Portland, Oregon, to assume his new duties as district ranger, the Gasquet Ranger District was the Siskiyou National Forest's southernmost district and the Pacific Northwest Region's only California district, and the Gasquet CCC camp was just being set up. Determining the Smith Fork Ranger Station—then located west of the current station—inadequate, Ranger Nielson decided the CCC camp's first project would be construction of a new ranger station. The regional office developed a site plan, and construction began.

While most of this historic compound's buildings—including the main office, an extensive 1938 remodeling by the CCC of an office built in the early 1930s—reflect the Pacific Northwest Region's standard Cascadian Rustic style, the ranger's residence was built in the Colonial Revival style popularized by the reconstruction of Colonial Williamsburg, Virginia, during the 1920s. Ranger Nielson's wife, it seems, convinced her husband that the ranger's residence should be built from a plan in the July 1930 issue of *Ladies Home Journal*. It was. Occupied by Ranger and Mrs. Nielson in 1933, the residence was the first structure completed by the CCC at the new ranger station. Since then, it has been the home of nine Gasquet district rangers and their families. Currently, it is used as the men's barracks.

When, in 1947, the Gasquet Ranger District was transferred from the Siskiyou National Forest to the new Six Rivers National Forest, the Gasquet Ranger Station buildings, built—with the exception of the ranger's residence—to Pacific Northwest Region specifications, became a Pacific Southwest Region architectural aberration. For that reason, this historic ranger station shows some design themes and motifs—the cut-out "pine tree logo" for example—not seen at other Forest Service administrative sites in California. And, thanks to Mrs. Nielson and the *Ladies Home Journal*, it has the first Forest Service structure built in the Colonial Revival style in California.

ACCESS: Historic Gasquet Ranger Station is located on U.S. Highway 199, the Smith River Scenic Byway, about 20 miles east of Crescent City, and the office is open during normal working hours.

Office building at historic Camptonville Ranger Station, Tahoe National Forest, California.

CAMPTONVILLE RANGER STATION
Tahoe National Forest, California
(1934)

Construction of Camptonville Ranger Station—now a Forest Service fire station—was just one of many changes witnessed by Frank W. Meggers during his 1927 to 1945 service as district ranger on the Tahoe National Forest's old Camptonville Ranger District.

Ranger Meggers, a U.S. Marine during World War I, went West to seek his fortune once the war was won. After harvesting wheat in South Dakota, working on ranches in Oregon and California, and laboring briefly as a millwright at Hobart Mills in California, he signed on with the Forest Service in 1920 as a fire guard at Camptonville under Ranger W.C. Whittum. Before long, he was assistant ranger at Camptonville, and then at Sierraville. He was appointed district ranger at Camptonville when Ranger Whittum died in 1927. In eighteen years as ranger at Camptonville, he saw the Forest Service replace horses with automobiles, iron telephones with radios, and—as a result of New Deal public works projects—his rented office over a Camptonville saloon with a new ranger station. As retired Ranger Meggers put it during a 1975 interview:

> They came along and said "If you had all the money you wanted, what could you do to help your district?" Of course, we wanted a headquarters building . . .

to replace that rented office over the saloon. As Ranger Meggers explained:

> We had our fire tools up there, mind you, and down the stairs we'd go if anything took place. It wasn't a proper setup, so through the CCCs we got this lovely setup, right in Camptonville.

Ranger Meggers's "lovely setup" was the historic Camptonville Ranger Station compound—an office, two residences, and two garages built to the California Region's standard Blanchard and Maher designs by the Civilian Conservation Corps between 1934 and 1936. These buildings—and other, non-historic buildings added later—were used as a ranger station until 1971, when the Tahoe National Forest combined the Camptonville and Downieville ranger districts. They continue in use as a Downieville Ranger District fire station.

Camptonville is no stranger to forest fires. It has been burned out at least twice since the Gold Rush days. And in 1959, when Lynn Horton was Camptonville district ranger, the town and the ranger station again lay in the path of flames. As the townsfolk prepared to evacuate, Ranger Horton's normally quiet Camptonville Ranger Station became the command post for a pitched battle. This time, directed by a team of Forest Service fire specialists, a force of bulldozers, pumpers, air tankers, and more than a thousand men fought desperately, stopped the fire at the highway, and saved Camptonville. In the town, people resumed their lives. And at Camptonville Ranger Station, Ranger Horton made plans to rehabilitate the gaping scar, twenty-five miles long by two miles wide, left by the fire.

Historic Camptonville Ranger Station, an excellent example of a 1930s ranger station compound, will continue in service as a Forest Service fire station. Also, the Tahoe National Forest and Camptonville have agreed to develop the residence as a museum of local and Forest Service history and an environmental education facility.

ACCESS: Historic Camptonville Ranger Station is located in the small town of Camptonville, just off California Highway 49 between Nevada City and Downieville.

WEAVERVILLE RANGER STATION
Shasta-Trinity National Forests, California
(1934)

For decades a steam whistle on the auto shop at Weaverville Ranger Station, for twenty years the old Trinity National Forest's headquarters, blew to let Weaverville's townsfolk know it was noon. Although the whistle is silent now, and a forest supervisor in Redding has overseen the combined Shasta-Trinity National Forests since 1954, the historic Forest Service compound has changed little since it was built at the northern end of the historic Gold Rush town in 1934.

The site on which Weaverville Ranger Station was built, at the confluence of three streams flowing out of the Trinity Alps to the Trinity River, was extensively placer mined in the late nineteenth century. This greatly changed the topography and stream courses. Sometime after World War I, probably in the early 1930s, the Forest Service acquired this property, a one-time jumble of town lots and mining claims. Then, with the Great Depression and its New Deal public works projects, came the opportunity to build a Trinity National Forest headquarters. And so, in 1934, with the $10,500 and labor furnished by the various New Deal relief agencies, the Forest Service built the ten-building compound that is today's Weaverville Ranger Station.

Weaverville Ranger Station, Shasta-Trinity National Forest, California.

Construction of the buildings, to the California Region's new standard Blanchard and Maher pre-cut designs for forest headquarters buildings, was supervised by three local Forest Service foremen. These men, Jim Everest and brothers Marvin and E.M. Goodyear, made Depression-era careers of supervising relief work projects. As then 92-year-old Marvin Goodyear told it in a 1990 interview, Trinity National Forest officials gave the three foremen the plans, the materials, the labor, and the go-ahead, and then "went back to their office and did not bother us again until it was completed."

During the summer of 1934, loads of ready-cut Douglas-fir, cedar, and redwood components—logged and milled not more than sixty-five miles away in Eureka, on the coast, but shipped some 260 miles south to San Francisco and trucked north to Weaverville because of poor roads over the Coast Ranges—arrived, and construction was under way. Given a free hand, the three foremen made design changes needed to fit the buildings to the site. The major

176

change they made was to place the ranger's residence at the west end rather than the east end of the site, and switch the ends of the office building around to meet this change. All this was done without consulting the architects who, on a trip from San Francisco to look the job over, told Mr. Goodyear that the office building was the most accurately built to their design in the region—even though they'd reversed the original design.

Much of the manual labor to build the compound was provided by unemployed local men. Leonard Morris, a Weaverville resident who worked on the project, recalled in a 1990 interview that, when men were needed for some job, he would take the Forest Service's Dodge fire truck out where they stayed, drive around blowing the siren, and men would come out of the woods ready to work. These men seem to have been employed for the Forest Service by the Works Progress Administration and National Recovery Act programs. Civilian Conservation Corps enrollees also worked on the project.

The compound, completed in 1934, comprised ten buildings. These included the headquarters office building, ranger's residence, auto shop, oil house, two warehouses, two large garages, and a storage building. In 1943, after California Highway 299 was routed past the front of the office and ranger's residence in 1939, a forest supervisor's residence was built across the highway from the compound. Of course, the uses of several of these buildings have changed with the compound's change in status from forest headquarters to ranger station and growth of the district ranger's staff. Both the ranger's and supervisor's residences, for example, now house offices.

As impressive as the buildings are the grounds. A CCC crew, based at a spike camp next to the compound, stabilized the three stream channels and carried out a landscape plan shortly after the buildings were completed. This landscape has matured beautifully over the decades.

This historic Forest Service compound, as both the Trinity National Forest headquarters and a Shasta-Trinity National Forests ranger station, has been an important part of historic Weaverville's community life for over six decades. It remains so today.

ACCESS: Weaverville Ranger Station is located on California Highway 299 at the northern end of Weaverville. Parking is available in front of the main office building and in a public parking lot across the highway.

DOUGLAS RANGER STATION
Sierra National Forest, California
(1934)

Roy Blood, a Sierra National Forest district ranger and deputy forest supervisor, knew what he wanted. And he got it. What he wanted and got was an official residence built at the old Douglas Ranger Station that can only be described as "a striking anomaly" among Forest Service buildings.

About the time the Sierra National Forest evolved from the Sierra Forest Reserve in 1908, the "old Douglas place" at which Ranger J.T. Nodden reported the "Government team [of horses and pack stock] used on the Forest" had been kept for some time was formally withdrawn for ranger station use. This original Douglas Ranger Station, a wood-frame house with a barn out back, served as a district ranger's headquarters through the early 1930s even though most Forest Service business was centered in North Fork a mile and a half up the road. Then, when the Civilian Conservation Corps built what became one of the largest Forest Service compounds in California at North Fork (pages 166 to 168), Douglas Ranger Station was finished . . . almost.

During that North Fork construction boom, Ranger Blood pressed for a new official residence—for himself—at the old Douglas Ranger Station site. The unusual result was construction in 1934 of the Douglas Station House to the direct specifications of the employee who was to live in it. And the building was as unusual as its origin. Blood insisted on living in something different from the standard ranger residence—something more akin to the houses popular in the San Francisco Bay Area he hailed from. So, instead of a standard Blanchard and Maher residence, Douglas Station House was built in the Cottage Revival style and looks like anything

178

Ranger Roy Blood's Douglas Station House at the old Douglas Ranger Station site, Sierra National Forest, California. U.S. Forest Service photo.

but a Forest Service building. While it would be at home in any California urban neighborhood, this stucco house looks out of place in its rural forested environment. The landscaping, which includes elaborate rockwork and exotic plants, adds to its anomalous appearance.

Ranger Blood's cottage at the old Douglas Ranger Station site has been lived in by Forest Service employees off and on since his years there. The Sierra National Forest plans to develop this unique Forest Service residence as a visitor and interpretive center at the beginning of the 100-mile Sierra Vista Scenic Byway.

ACCESS: *Douglas Station House is located about 1/2 mile east of North Fork on Forest Road 7S09, Douglas Station Road, and about 35 miles north of Fresno. Until opened as a visitor center, it may be viewed from the outside.*

Historic Placer Guard Station, Sierra National Forest, California. U.S. Forest Service photo.

PLACER GUARD STATION
Sierra National Forest, California
(1936)

When fire guard Jesse A. Love and his wife arrived for the summer of 1936, they were not impressed with Placer Guard Station. The site, overlooking a meadow along the banks of Chiquito Creek, was beautiful. But the rustic cabin of vertical plank siding and shake roof, built sometime around World War I by an early fire guard, was ramshackle and dirty. Minarets District Ranger Ed Madsen gave the Loves permission to make any improvements to the building they wished. After fixing the place up a bit, they tore the cabin down and built a new one.

Not long after their arrival, the Loves acquired a supply of lumber salvaged from the kitchen building of an abandoned Civilian Conservation Corps camp at nearby Logan Meadow. With the help of CCC foreman and carpenter Al Rayden and three CCC enrollees,

Love tore down the old cabin and built the new guard station residence on the same site. Working together, the five men completed the new house in about ten days. It was covered with narrow shakes left as scrap from cutting regular shakes. While the new cabin was being built, the Loves lived in a tent in the meadow.

After the new house was completed, Love developed a spring north of the cabin as a water supply. He built and installed a wooden tank and copper pipeline from materials once used by bootleggers whose illegal still near Friant was confiscated by federal agents. The Loves used a propane stove for cooking, gas lanterns for light, and a wood burning stove for warmth and to heat water.

As a fire guard, Love was on call throughout fire season. When he wasn't on a fire, he was required to work eight hours a day. Love used that time well. He not only built the residence, but also the storage building, garage, and outhouse as well as extensive rock retaining walls, completing most of the work during the first two of his ten seasons at Placer Guard Station. To build the rock walls, he hauled sand and rock in his own pickup truck from Soda Springs Campground, widening the trail by hand to allow his truck to pass. And he did all this in addition to fighting fires and maintaining six campgrounds.

Placer Guard Station, a facility built of mostly salvaged and second-hand materials, was used by Forest Service guards for decades after Jesse Love's time there. It remains a monument to one fire guard's ingenuity and hard work.

ACCESS: Placer Guard Station is about 35 miles from North Fork and just off paved Forest Road 4S81, the Sierra Vista Scenic Byway. From that road, take Forest Road 6S25 which winds southeasterly for about a mile to its junction with Forest Road 6S61Y. Turn right onto this road and follow it west for about mile to the station.

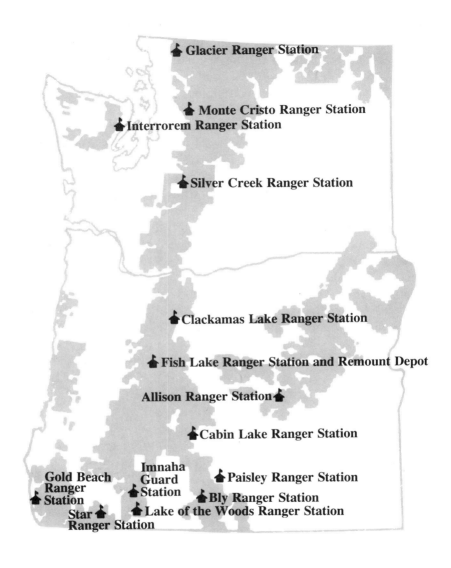

Fourteen historic U.S. Forest Service region stations of the Pacific Northwest Region (Region 6) in Washington and Oregon.

182

Chapter Six
PACIFIC NORTHWEST REGION

Nineteen national forests—six in Washington and thirteen in Oregon—totaling more than twenty-four million acres make up the Pacific Northwest Region, or Region 6. Bisected north to south by the Cascade Ranges, the region's wet western side and dry eastern side have some of the most diverse terrain, climate, and vegetation found in the United States.

Along or near the Pacific Ocean, three of the region's national forests include Olympic Peninsula rainforests, rugged Oregon coastlines and sand dunes, the Douglas-fir and western hemlock forests of the Coast Ranges, and the mixed coniferous-broadleaf forests of the Klamath Mountains. Inland, an almost unbroken chain of national forests occupies both flanks of the majestic Cascade Ranges from Canada to California. The great High Cascades volcanoes, dominating the landscape and determining the climates, separate the dense Douglas-fir forests of the western slopes from the dryer eastside pine forests. Farther east, the pine forests of the Blue Mountains of northeastern Oregon and northeastern Washington's scrap of the Rocky Mountains round out Region 6's national forests. A regional forester in Portland, nineteen forest supervisors, and about eighty district rangers manage these national forests that, until recently, produced about half the timber sold annually from the National Forest System, and continue to provide a wide range of benefits.

The fourteen historic ranger and guard stations featured in this chapter span the first three decades of Forest Service history as well as the diverse national forest environments of the Pacific Northwest Region.

183

INTERROREM RANGER STATION
Olympic National Forest, Washington
(1907)

The old Interrorem Ranger Station cabin on the Duckabush River, also known in the past as Duckabush Ranger Station and more recently as Interrorem Guard Station and Interrorem Cabin, was the first ranger station on the Olympic National Forest which, at the time, included the area that is now Olympic National Park. In addition to being the oldest Forest Service building on the Olympic Peninsula, this cabin—in an old orchard surrounded by a damp Douglas-fir forest—is one of the oldest remaining Forest Service ranger stations in the Pacific Northwest Region.

Ranger Emery J. Finch, a son of Hoodsport pioneers, built the cabin in 1907 as authorized by Forest Supervisor Fred Hanson, first Olympic National Forest supervisor. Ranger Finch and his new bride, Maybell Peterson, first occupied the new Interrorem Ranger Station's square, one-story, three-room, peeled-log cabin set off by a long front porch and a pyramidal cedar shake roof on April 22, 1908.

From 1909 to 1933, the station served as an administrative site for both the Olympic National Forest and the Forest Service-administered Mount Olympus National Monument that later became Olympic National Park. At times during the early 1920s, when this apparently "interim" station was not used to house Forest Service personnel, it was used occasionally by the public for birthday celebrations, family reunions, and similar events. From 1933, the year the nearby Mt. Jupiter fire lookout was built, to 1942, the Interrorem station hosted several Depression-era government programs including the Works Progress Administration and the Civilian Conservation Corps even as it served as a fire guard base.

Forest Service fire guards continued to be based at Interrorem Guard Station during World War II and for four decades after the war. Between the years 1947 and 1952, when Merritt B. Major was fire guard there, the telephone was replaced by two-way radios and electric lights were installed at the station. And there was a garage, probably built during the 1930s and now gone. Beginning in 1986, the Interrorem Cabin was used by Forest Service volunteers, and

184

Historic Interrorem Ranger Station, Olympic National Forest, Washington.

Interrorem Ranger Station under construction in 1907 (above), and Interrorem Ranger Station cabin and garage in 1937 (below). U.S. Forest Service photos.

in 1994 it became an Olympic National Forest recreation rental cabin. Visitors who rent the cabin can experience something of early forest ranger life—without, for example, electricity or running water.

ACCESS: To reach historic Interrorem Ranger Station, turn off U.S. Highway 101 about 2 1/2 miles south of Brinnon and 22 miles north of Hoodsport onto paved Forest Road 2510, Duckabush Road, and follow the road for 4 miles to the end of the pavement. A parking lot and Interrorem Cabin are on the left. Contact Hood Canal Ranger District, P.O. Box 68, Hoodsport, Washington 98548, telephone (360) 877-5254, for Interrorem Cabin rental information.

ALLISON RANGER STATION
Ochoco National Forest, Oregon
(1911)

The story of historic Allison Ranger Station—now Allison Guard Station—has a beginning and an end, but part of the middle is missing.

Ranger E.W. "Cy" Donnelly, the first ranger of the Snow Mountain Ranger District, Ochoco National Forest, built the first Allison Ranger Station cabin, the district's first headquarters, in 1911. About 200,000 acres of national forest lands were administered from this one-room, pine-log cabin, also known as the Donnelly Cabin.

Then, about 1925, another cabin seems to have been built, either to supplement or replace the Donnelly Cabin. Only photographs remain of this somewhat larger cabin and two smaller outbuildings that seem to have been used for less than ten years. When, how, and why these structures, located between Ranger Donnelly's 1911 cabin and a new ranger station complex built in 1935, were built and disappeared is mostly unknown. One of the outbuildings, sold at auction to a rancher, was moved to a location near Paulina, about twenty-five miles to the northwest.

Ranger Cy Donnelley's restored 1911 Allison Ranger Station cabin, Ochoco National Forest, Oregon.

The current Allison Guard Station complex, seven of the eight Allison Ranger Station buildings constructed by the Civilian Conservation Corps in 1935 about 200 yards southwest of Ranger Donnelly's cabin, served as the Snow Mountain District's ranger station until the current ranger station compound in Hines was constructed in the 1950s. As a result, two historic ranger stations—the one built in 1911 and the other in 1935—remain at Allison.

Ranger Donnelly would have preferred the name Howard Ranger Station for the general who pursued Chief Egan and his band of Paiutes, along with some Shoshone and Bannock Indians, northward through nearby Howard Valley in 1878. But the station took its name from the nearby Jesse Allison homestead. In addition to the cabin, the original Allison Ranger Station included a tool storage shed, outhouse, and corral.

Ranger Donnelly's one-room log cabin was restored in 1993 by Snow Mountain Ranger District personnel, some of whom donated their own time, with help from the Pacific Northwest Region's historic preservation team, Passport in Time volunteers, Youth

The office-cookhouse (above) and warehouse (below) at Allison Ranger Station, Ochoco National Forest, Oregon, in the 1930s.'U.S. Forest Service photos.

189

Conservation Corps enrollees, Native American students with the Youth Opportunities Program of the Burns Paiute Tribe, and others. The cabin was stabilized and raised to remove and replace ant-eaten and decayed sill logs, a gravel pad was built to enclose the foundation, new floor joists and boards were put in place, and porch support posts were set. Larch shakes, split with wooden mallet and froe, provided a new roof. The logs were coated with preservative. An archaeological study preceded this restoration work. Thanks to this effort, visitors may see and appreciate what it was like to live and work on the Ochoco National Forest during the early days.

Life and work continue at Allison Guard Station, where the CCC-constructed buildings provide a primary base for Forest Service fire and project crews. These buildings, built in the Pacific Northwest Region's distinctive Depression-era Cascadian Rustic style, include an office-cookhouse building, a warehouse, and residences.

Plans call for development of a visitor's trail along which both the 1911 cabin and the 1935 compound will be interpreted.

ACCESS: Current access to historic Allison Ranger Station is by appointment with the Snow Mountain Ranger District in Hines, telephone (503) 573-7292. There are two primary ways to reach Allison Guard Station from U.S. Highway 20. At Hines, turn off the highway onto Forest Road 47 opposite the Tecton mill, and follow this road to its junction with Forest Road 41. Bear left onto Forest Road 41, and follow it into the Ochoco National Forest and past Delintment Lake to Allison Guard Station. Most of this approximately 45-mile route is paved one-lane road with frequent turnouts. Or, about a mile west of Riley, turn north off U.S. Highway 20 and follow Forest Road 45 for 35 miles, turn left onto Forest Road 41 and continue for 3 miles to Allison Guard Station.

Historic Star Ranger Station, Rogue River National Forest, Oregon.

STAR RANGER STATION
Rogue River National Forest, Oregon
(1911)

"Pretty humble" is the way Jeff LaLande, Rogue River National Forest archaeologist, described the original Star Ranger Station, "one of the oldest remaining Forest Service structures in the nation still in use." Southeast of Medford in the rural, agricultural valley of the Applegate River, it's within sight of the crest of the Siskiyou Mountains.

This building, a 12-foot by 20-foot, wood-frame structure with shiplap siding and a wood shingle roof, was built in 1911 by two local carpenters hired by the Forest Service to build a ranger station. As such, it was meant to be used as both a residence and an office by the Applegate District ranger. That ranger used it as an office, but preferred to live in a tent until a large bungalow built at the ranger station in 1914 provided both a better office and living quarters. After 1914, the 1911 building was used to store horse tack, and became known as the "old tack room" around the station. Later still, it was used to store tree marking paint and other supplies. It's been moved to at least three locations within the Star Ranger Station compound over the years.

Multi-pane, double-hung sash windows and knee-braces for the eaves are about the only features that keep this building from looking like an old backyard shed. And, as far as looks go, this building reflects the fact that many of the earliest Forest Service structures were not built to "look rustic," as they were during Civilian Conservation Corps days a couple decades later, but to be functional at the lowest possible cost. Had logs been readily available, it likely would have been a log cabin.

Star Ranger Station was home to Ranger Lee C. Port from World War I to World War II. An amiable but no-nonsense ranger during his many years on the Applegate District, the Oklahoma-born ranger was readily identifiable at a long distance when riding the high country meadows because of his white horse. His horse's tack, presumably, was kept in the 1911 building. Ranger Port, popular with local folks, recorded some of the Applegate country's early history before the old-timers he knew passed away.

That history, as well as local Forest Service history, will be interpreted in the 1911 Star Ranger Station building that is being restored inside and out and prepared for that purpose. An interpretive sign outside the building is the first evidence of this plan. The inside exhibits remain in the future.

ACCESS: The historic 1911 Star Ranger Station building is at the north end of the public parking area at Star Ranger Station, about 20 miles by paved road from Medford and Interstate Highway 5. From Medford, take Oregon Highway 238, also called West Main Street in Medford and Jacksonville Highway beyond, through the historic mining town and former county seat of Jacksonville, and continue another 8 miles to the community of Ruch. Turn south onto Upper Applegate Road where road signs point to Applegate Lake and Valley View Vineyards, and drive another 6 miles to Star Ranger Station.

FISH LAKE RANGER STATION & REMOUNT DEPOT
Willamette National Forest, Oregon
(1921)

The old Fish Lake Ranger Station, now a Willamette National Forest work center comprising the Fish Lake Guard Station and Remount Depot, dates from 1906 when its twenty-acre site on Fish Lake's northeastern shore was withdrawn from potential homestead entry for Forest Service administrative use. During the previous four decades, from 1868 to 1906, a way station for Santiam Wagon Road travelers occupied the site.

A log cabin ranger station was built in 1908 along the old Santiam Wagon Road to support fire patrols in the surrounding forest. The station began service as the old Santiam National Forest's summer field headquarters in 1911. The original cabin was crushed by snow during the winter of 1912-13. By the summer of 1914, the Cascadia Ranger District of the Santiam National Forest had replaced the cabin with a 14-foot by 28-foot two-room cabin and built a 26-foot by 28-foot six-stall barn at a total cost of $436.74. An 800-foot wire fence added $51.74 to the price of this ranger station.

As time went on, other buildings were added. The cabins still in use include the dispatcher's cabin built in 1921, the supervisor's cabin built in 1924 to house Santiam National Forest supervisor C.C. Hall (so it's called "Hall House"), and the commissary cabin built in 1924. Also remaining are the springhouse shed next to the dispatcher's cabin and a fire hose shelter, also built in 1924. In 1926, the station was described as "a group of very attractive log cabins . . . [that] house the [fire] dispatcher and his office, the two or three firemen stationed here, [and the] packer and pack animals."

Fish Lake itself, while filled with water in the winter and spring, is dry during the summer. Its dry bed was used to grow hay for the pack animals. Hackleman Creek, which flows into Fish Lake from the west, harbors the lake's fish when it drains naturally through its porous lava-rock bottom.

During the 1930s, Fish Lake became an important fire fighting remount station for crews and pack animals sent to forest fires throughout the central Cascades, and served as a Civilian Conser-

The 1921 dispatcher's office cabin at Fish Lake Ranger Station on the old Santiam National Forest in 1926 (above, photo courtesy of Dorothy Lueck and Ron Johnson) is still in service at Fish Lake Guard Station and Remount Depot (below), Willamette National Forest, Oregon.

vation Corps work camp on the Willamette National Forest—a forest formed in 1933 by merging the Santiam and Cascade national forests. Beginning in 1934, the CCC built several buildings and other facilities at the site. By 1940 the station comprised just over a dozen major structures.

Fish Lake remained a fire fighting remount depot into the 1960s, when vehicles and aircraft largely replaced pack animal transportation of crews and equipment to forest fires. Today the historic 1921 dispatcher's cabin and 1924 commissary cabin house the Fish Lake Guard Station. The combination log and frame 1924 Hall House houses the Willamette National Forest's packer. And the barns, sheds, corrals, and lava rock walls built by the CCC in 1934 serve as the Fish Lake Remount Depot that provides pack animals to support wilderness, trails, and fire management operations. A crew barracks built in 1960 houses employees.

In more recent years, active and retired Forest Service personnel and others have worked to preserve Fish Lake's still-serviceable 1920s-vintage log cabins and to carry on the Forest Service tradition of supporting backcountry work with horses and mules.

ACCESS: To visit Fish Lake Guard Station and Remount Depot, turn off Oregon Highway 126, the McKenzie Highway, 2 miles north of Clear Lake and about a mile south of U.S. Highway 20, the South Santiam Highway, into the Fish Lake Campground. Drive about 50 yards north to the cul-de-sac marked by "No Camping Here" signs, park, and walk northwest along the old Santiam Wagon Road, passing through two gates. The remount depot barns are on the right, the dispatcher's cabin and commissary cabin of the guard station are straight ahead beyond the flag pole, and Hall House is beyond them and to the left.

Ranger's residence at historic Cabin Lake Ranger Station, Deschutes National Forest, Oregon.

CABIN LAKE RANGER STATION
Deschutes National Forest, Oregon
(1923)

Forty miles south-southeast of Bend, Oregon, remote Cabin Lake Ranger Station—now a guard station used by seasonal fire and project crews—was headquarters for five Deschutes National Forest district rangers between 1921 and 1945. Except for a bunkhouse and a pump house built in 1923, all the remaining Cabin Lake buildings were constructed by the Civilian Conservation Corps between 1934 and 1938.

Cabin Lake Ranger Station wasn't the Fort Rock Ranger District's first ranger station. That first district headquarters was located in a small house in the Fort Rock Basin town of Fort Rock by Ranger W.O. Harriman—later Deschutes National Forest assistant supervisor and Ochoco National Forest supervisor—in 1914. At that time, Fort Rock—about eight miles south of Cabin Lake and about two miles south of Fort Rock, the massive volcanic tuff ring from which the town and the ranger district take their names— was quite a settlement. Before long, Ranger Harriman rented an old Fort Rock stage station and barn and moved his headquarters to them. But Ranger Harriman didn't live there. Instead, he took up a homestead about two miles south of Cabin Lake and lived there in a tent house and shack. His district headquarters remained in Fort Rock until 1921 when his successor, Ranger J. Roy Mitchell, moved it to Cabin Lake.

A well drilled at Cabin Lake in 1916 to improve grazing on Fort Rock District range made the site, on the edge of the national forest and overlooking the Fort Rock Basin where the ranchers who grazed the district's range lived, an attractive location for a ranger station. By the time Ranger Henry R. Tonseth took over the district in 1934, the CCC was replacing most of the original ranger station buildings at Cabin Lake with a ranger station compound that looked like it was there to stay. Six of the seven structures they built between 1934 and 1938—two residences, a cook house, a warehouse, a shop, and a gashouse—remain among the widely-spaced ponderosa pines at the edge of the forest. Sturdy wood-frame buildings on poured concrete foundations, all have clapboard siding

Warehouse at historic Cabin Lake Ranger Station, Deschutes National Forest, Oregon.

and wood-shingle, medium-gable roofs. Only the warehouse sports the "pine tree logo" found on many CCC-constructed ranger station buildings. Another CCC building, a bunkhouse, was moved about forty-five miles west to Crescent Lake Ranger Station in 1949.

During the late 1930s and early 1940s, Cabin Lake Ranger Station was the hub of most Deschutes National Forest timber sale activity. Forest Service personnel administering sales to the Brooks-Scanlon and Shevlin-Hixon mills in Bend lived there.

Ranger Tonseth, whose almost thirty-five years as Fort Rock District ranger established a Pacific Northwest Region record for service on one district, was the only district ranger to reside at the CCC-built Cabin Lake Ranger Station. In 1945, not quite a third of the way through his Fort Rock District service, his headquarters were moved from Cabin Lake to Bend. A well-liked, hard-working, dedicated Forest Service officer, Ranger Tonseth retired in 1968.

And where is the lake for which Cabin Lake Ranger Station is named? It's about an eighth of a mile south of the station and east of Forest Road 18. But don't take your boat! Cabin Lake is a small, dry lake that holds water in only the wettest years.

ACCESS: To reach historic Cabin Lake Ranger Station, turn east off U.S. Highway 97 at LaPine, about 30 miles south of Bend, onto paved Forest Road 22. Drive about 26 miles east to Forest Road 22's junction with Forest Road 18, just over a mile east of South Ice Cave, turn south on Forest Road 18 and follow it for just over 6 miles to Cabin Lake Guard Station on the right.

SILVER CREEK RANGER STATION
Mt. Baker-Snoqualmie National Forest, Washington
(1931)

One of only three log office buildings known to have been built at Pacific Northwest Region ranger stations during the Great Depression is the centerpiece of the Silver Creek Guard Station Historic Site. Nestled in an old-growth forest of Douglas-fir and Pacific silver fir along the Mather Memorial Parkway, which is Washington Highway 410, the historic log office building is just north of the White River Entrance to Mount Rainier National Park.

Silver Creek Ranger Station was established just after World War I when completion of a road to the locality and the growing popularity of the automobile made a Sunday trip from Seattle, Tacoma, and other nearby towns to Silver Creek and on to Mount Rainier National Park a popular outing. The ranger station, staffed during the summer only, was conveniently situated to administer the Forest Service's share of this growing recreational use; winter work was carried out from the old Snoqualmie National Forest's headquarters in Tacoma.

The original facilities at Silver Creek Ranger Station were a frame building that housed an office, living quarters, and storage space, and a three-sided garage-woodshed. A residence and some outbuildings were added during the 1920s. When the main structure burned to the ground in 1930, that new log office building—a Forest Service rarity in that time and place—was built the following year.

Historic Silver Creek Ranger Station, Mt. Baker-Snoqualmie National Forest, Washington. U.S. Forest Service photo.

A log addition was added in 1936. The fact that the ranger station was surrounded by one of the largest groups of log cabins in the United States—built, many of them in the early 1920s, on national forest summer home plots—may explain why the office was built of logs. Nevan McCullough, ranger at the time of construction, attributed the log building to Olie Strom and Peter Brien, whom he said "were exceptionally good workmen, taking pride in their joints."

Today, the well-preserved log office building serves as a Silver Creek Guard Station bunkhouse.

ACCESS: The historic Silver Creek Ranger Station log office building is located on the western side of Washington Highway 410, the Mather Memorial Parkway, about 1/4 mile north of the Mount Rainier National Park boundary.

Protection assistant's residence at historic Clackamas Lake Ranger Station, Mt. Hood National Forest, Oregon.

CLACKAMAS LAKE RANGER STATION
Mt. Hood National Forest, Oregon
(1934)

The first Clackamas Lake Ranger Station structures, a log cabin and a barn built by pioneer ranger Joe Graham in 1906, are long gone. But eleven of their twelve successors, wood-frame buildings beautifully crafted on the same site during the Great Depression, remain as Clackamas Lake Historic Ranger Station.

By the time Ranger Graham finished his 24-year tour as district ranger at Clackamas Lake in 1930, forest use and management had increased to require a larger administrative complex. A twelve-building ranger station compound design was completed in 1932. Establishment of the Civilian Conservation Corps in 1933 and availability of New Deal funds allowed rapid completion of this compound during the next two years. And, since the idea was to employ out-of-work men, the original conservative plans were modified to include labor-intensive items that kept the men employed longer.

Smokey Bear aboard the 1935 Chevrolet fire truck based at historic Clackamas Lake Ranger Station, Mt. Hood National Forest, Oregon. Photo by Pete Martin.

While skilled carpenters working for the Forest Service did most of the actual building construction, the CCC men did most of the rock work and landscaping. The result was one of the Pacific Northwest Region's more beautiful Depression-era ranger stations.

Clackamas Lake Ranger Station, named for six-acre Clackamas Lake near the headwaters of the Oak Grove Fork of the Clackamas River, has served continuously as a Forest Service administrative site since 1906. Nestled in a Douglas-fir forest at an elevation of 3,500 feet where six-foot snows are not uncommon, it was the summer headquarters of the Clackamas Lake Ranger District—first on the old Oregon National Forest and then on the Mt. Hood National Forest—until 1952. Today, although still in use as a summer guard station, work center, and employee residence, Clackamas Lake Ranger Station looks about as it did in the 1930s. Its access road was paved in 1962, but no modern buildings have intruded.

A group called Friends of Clackamas Lake Historic Ranger Station helps the Forest Service preserve and interpret this national historic site. A visitor and interpretive center is housed in the ranger station office building, restored and refurnished to look like a 1930s ranger's office. Experienced volunteers offer guided tours of the ranger station compound—residences, warehouses, a barn, a blacksmith shop, and other buildings—and other services to visitors.

An historic 1935 Chevrolet fire truck, donated by Friends of Clackamas Lake Historic Ranger Station in 1993 and restored as a fire prevention "vehicle" for Smokey Bear's fiftieth anniversary in 1994, is based at Clackamas Lake Ranger Station. The truck visited over twenty fire prevention events in Oregon and Washington during 1994, and continues to support fire prevention efforts.

ACCESS: Historic Clackamas Lake Ranger Station, about 75 miles southeast of Portland, is open to visitors from 9:00 a.m. to 5:00 p.m., Wednesday through Sunday, from July 4 through Labor Day weekend, and on weekends only from Memorial Day weekend to July 4. To get there, take U.S. Highway 26 to its junction with Skyline Road, which is Forest Road 42, about 11 miles east of Government Camp and 5 miles west of the Warm Springs Indian Reservation-Mt. Hood National Forest boundary. Turn south onto Skyline Road, and follow this paved road 9 miles to the ranger station.

GOLD BEACH RANGER STATION
Siskiyou National Forest, Oregon
(1936)

On the wind-swept coast of southwestern Oregon, where the ancient Klamath Mountains meet the mighty Pacific Ocean, Gold Beach Ranger Station has braved the elements and served the public since the Civilian Conservation Corps built it for the Forest Service in 1936. Today, it remains not only the picturesque headquarters of the Gold Beach Ranger District, but also a vital part of the Gold Beach community.

Although only one of many Depression-era ranger stations still in service in the Pacific Northwest Region, Gold Beach Ranger Station is an outstanding example of the planned Forest Service

Gold Beach Ranger Station, Siskiyou National Forest, Oregon, in 1940. U.S. Forest Service photo.

administrative compound of that time. Arranged on five levels, its original nine Cascadian Rustic style buildings are at once aesthetically united and functionally separated on a sloping marine terrace site purchased by the government for $2,200 in 1935.

After necessary terracing and grading of that site, construction of those nine Gold Beach Ranger Station buildings was begun in January 1936 by a CCC crew from Gasquet, California, about ten miles south of the state line. One of their first jobs was quarrying stone for retaining walls and decorative features on the buildings. In April, a crew of twenty-four carpenters from an Ohio-recruited CCC crew joined the effort. These crews worked quickly, and the new ranger station soon took shape. By late February 1937 Siskiyou National Forest employees had begun moving from their old office just south of the Curry County Courthouse in downtown Gold Beach, to the new ranger station.

As specified in the site plan, the Gold Beach Ranger Station office building was built at the entrance to the station where it served as a control point for all business traffic and was readily accessible to the public. Behind the office were the residential and service areas.

Three residences were built on the terraced slopes east of the office. The ranger's residence, on the uppermost terrace, had the most privacy. Late in the summer of 1937, the Siskiyou National Forest supervisor and his other district rangers met in Gold Beach to help District Ranger Ed Marshall put the finishing touches on this house. The protective assistant's residence was located just above the office, where there was a clear view of the entry road and where travelers could reach it easily after hours. The fireman's cabin was on the opposite side of the entry road. Separate from this group of residences, the crew house was located between the office and the service court.

The service court, on a broad graded terrace above and northeast of the office, included a warehouse, machine shop, equipment storage building, and gas and oil house. Gold Beach's often severe winds may not have been fully considered when the ranger station was planned. As sited, with nothing to deflect the direct force of the wind, the equipment storage building was and remains vulnerable to high winds. Its original side-opening doors could not withstand the wind pressure, and vehicles had to be parked abutting the doors on the interior to prevent them from blowing inward. In fact, during the severe Columbus Day storm in 1962, the building was lifted from its footing by the winds.

As harmonious a group of government buildings to be found anywhere, all the original Gold Beach Ranger Station structures have horizontal clapboard siding, board and batten gables, and wood-shingled roofs. Accent masonry of distinctive beige rock, squared and coursed, is featured in foundations, chimneys, walks and porches. And, as in many CCC structures, the distinctive "pine tree logo" appears on the buildings' gables. Decorative iron work, including pine cone door knockers and tree-shaped hinges, add to the effect.

This may seem excessive for a government project. But the price of the land and the original cost estimates for the nine buildings—built by dollar-a-day CCC enrollees—came to less than $20,000. Even if labor costs weren't included in that figure, that wasn't a high price to pay for a nine-structure ranger station compound now nearing its seventh decade of service—a project that made good use of the exceptional masonry, carpentry, iron work, and landscaping skills of otherwise unemployed and discouraged young men.

The landscape plan for Gold Beach Ranger Station required many stone retaining walls—of both mortared and dry masonry—to contain the terraces, as well as other built landscaping features including mortared stone tree wells, stone curbs, and stone steps. Planted native trees, shrubs, and ground covers set off lawns that surround the office, the residences, and the crew house. According to George Morey, fire control officer at Gold Beach Ranger Station in the 1950s and 1960s, the ranger station

> . . . compound as we see it today looks much different than it did back in those days. The site was in a natural condition. . . . The brush and trees were at least thirty feet tall. The ranger's house was not visible from the office. An adult could not penetrate the jungle. Several families of raccoons lived in it. At night they would come down to the house and argue with the dogs and cats about the food. We managed to live with them, although a lot of sleep was lost at times. We did get even with one old coon. . . . The ranger's wife cooked it, and really it wasn't so bad.

The "jungle" of the station's landscape changed with the spring, 1954, arrival of new District Ranger Dan Abraham. As Mr. Morey recalled in 1986:

> Up to this time it bordered upon a crime to even think of lopping a branch from one of the bushes or trees. So with lightning speed, even before the environmental watchdogs knew what was happening, the entire crew was put to clearing the land—and clear it they did.

More recently, efforts to restore the ranger station's original landscape have uncovered CCC-built rock walls overgrown for decades.

Along with the landscape, use of some Gold Beach Ranger Station buildings has changed over the years. The office building still houses the district ranger's office, but the protective assistant's residence and crew house have been converted to office use and the shop building houses the fire management staff. An intrusive,

District ranger's office (above) and residence (below), Gold Beach Ranger Station, Siskiyou National Forest, Oregon.

two-story office building, built in 1964 just north of the original office building, provides still more office space. The district ranger still lives in the ranger's residence, and the fireman's cabin remains an employee residence.

There's more to Gold Beach Ranger Station, of course, than wood and stone. It has, for six decades, been an important part of this coastal Oregon community at the mouth of the wild and scenic Rogue River. Of course, the district ranger based there has managed many of the resources—including the increasingly important recreation resource—vital to the town's economy. But there's more to it than that—shared experiences that, down through the years, have forged a bond between the ranger station and the town. "The Forest Service telephone system," for example, as Mr. Morey recalled,

> served the public well, especially between Gold Beach and Agness. The telephone company had a line to Agness, but it never worked in wet weather, so much private business was done over the government lines.

Gold Beach Ranger Station hasn't always gone by that name. From the time it was built in 1936 until 1945, it was Chetco Ranger Station and headquarters of the Chetco Ranger District. When, after World War II, the Siskiyou National Forest reorganized, its Agness Ranger District was eliminated. Its old Agness Ranger Station, about thirty-five miles up the Rogue River from Gold Beach, was closed, and its lands incorporated into other districts with the new Gold Beach District getting the lion's share. A new Chetco District, headquartered at Brookings, extends from its Pistol River boundary with the Gold Beach District to the California line.

In addition to the historic Gold Beach Ranger Station compound itself, visitors may see many items of historic interest just inside the main entrance of the original office building—just to the right of the flag pole—including a Forest Service telephone switchboard that dates from the 1930s, weather recording devices, myrtlewood furniture hand-built for ranger stations, and photographs.

ACCESS: Gold Beach Ranger Station is located on the eastern side of U.S. Highway 101 in southern Gold Beach, and is open throughout the year during normal office hours.

Historic Monte Cristo Ranger Station, now Verlot Public Service Center, Mt. Baker-Snoqualmie National Forest, Washington.

MONTE CRISTO RANGER STATION
Mt. Baker-Snoqualmie National Forest, Washington
(1936)

The old Monte Cristo Ranger Station was headquarters of the old Mt. Baker National Forest's Monte Cristo Ranger District from 1936 when its Civilian Conservation Corps builders finished the station's office and residence until 1982 when the district was combined with the Darrington Ranger District and headquarters were moved to Darrington. At some time during this 46-year period, Monte Cristo Ranger Station was called Verlot Ranger Station.

Now called Verlot Public Service Center, the historic ranger station provides information services to several thousand Mt. Baker-Snoqualmie National Forest visitors each year. Additionally, the center houses an historical museum opened in 1989. Developed by a partnership between the Forest Service, Snohomish County

CCC alumni, and local volunteers, the museum interprets aspects of the South Fork Stillaquamish River Valley's history.

ACCESS: Historic Monte Cristo Ranger Station is located on Washington Highway 92 about 10 miles east of Granite Falls. Operated as the Verlot Public Service Center, it is open during the summer from 8:00 a.m. to 5:00 p.m., Monday through Thursday, and from 8:00 a.m. to 6:00 p.m., Friday through Sunday.

BLY RANGER STATION
Fremont National Forest, Oregon
(1936)

A rustic masterpiece in wood and stone, the Bly Ranger Station compound in Bly, Oregon, was built under U.S. Forest Service supervision by Civilian Conservation Corps enrollees and local experienced men between 1936 and 1942. Over a half century later, this group of seven administrative and residential buildings—augmented by a 1960s office structure—remains in service as headquarters of the Bly Ranger District.

As the Roosevelt administration's so-called "second New Deal" got under way in 1935, a four-acre ranger station site in Bly was acquired from Anna Avery for $625 in emergency relief funds. Before long, a team of experienced local workers and as many as 250 CCC workers stationed at Camp Bly were busy building the new ranger station as well as working on other jobs. By 1942, when the CCC was disestablished and Camp Bly was closed, a new district office building, new houses for Ranger Perry Smith and his assistant ranger, a guard residence, a garage, a warehouse, and a gas and oil house graced the sere Bly landscape. All these Cascadian Rustic style buildings were of native stone and lumber. Instead of the usual rough shake shingles associated with this style, the Bly buildings had sawn shingles. In addition to the "pine tree logo" on the buildings' gables and shutters, a pine tree symbol of green stone, incorporated in the stonework, gave the buildings a character all

District ranger's residence (above) and former office (below), Bly
Ranger Station, Fremont National Forest, Oregon.

211

their own. A 400-foot stone wall separated the compound from the highway.

Over the years, as use of some buildings has changed, the compound has aged gracefully. As the district's staff grew, more office space was needed. A new district office building was added in the 1960s, and the original office as well as the guard residence and the garage are used for office space. But the two main residences still house the district ranger and other Forest Service personnel, and the warehouse and gas and oil house continue to serve their original purposes. The entire compound is well preserved and maintained.

ACCESS: Bly Ranger Station is located on the south side of Oregon Highway 140 in the town of Bly, 53 miles east of Klamath Falls and 43 miles west of Lakeview. A self-guided tour folder, available at the Bly Ranger Station office, helps visitors tour the compound and appreciate its history as well as its current operations.

IMNAHA GUARD STATION
Rogue River National Forest, Oregon
(1936)

Almost any Oregonian would expect to find Imnaha Guard Station in northeastern Oregon, where the Imnaha River flows from the Wallowa Mountains into the Snake River in Hells Canyon, not on the western slope of the Cascade Range in southwestern Oregon. But that's where it is, tucked in the Douglas-fir forest about twenty miles southwest of Crater Lake. It seems an early Butte Falls area settler, fresh from northeastern Oregon, brought the name Imnaha and others with him, applied them to area creeks, and they stuck.

Imnaha Guard Station, a classic Cascadian Rustic style residence and garage built by the Civilian Conservation Corps between 1935 and 1937, replaced an old Forest Service cabin built on the then-Crater National Forest site between 1908 and 1910. Known as "Imnaha Tool Cabin," it was a simple, two-room, shake-over-pole or log structure, built as a tool cache and occasional shelter

Imnaha Guard Station, Rogue River National Forest, Oregon, in 1936, as roof was being shingled. U.S. Forest Service photo.

for seasonal trail and fire crews. In 1911, the small meadow at Imnaha was fenced to provide forage for Forest Service pack stock. This original Imnaha Guard Station, near major springs next to a lush meadow at the junction of several major trails, was no doubt a popular camping spot for Forest Service personnel and local area residents. It was reached only by trail until sometime between 1925 and 1930 when the road from Prospect was built.

During the New Deal years, the CCC built not only the new Imnaha Guard Station buildings but also the adjacent Imnaha Campground. During the post-War years, improved access made Imnaha Campground popular, especially during hot weekends in the Rogue Valley and hunting season. The campground was improved and expanded during the 1950s and 1960s, and the guard station continued to house seasonal Forest Service employees. The residence—complete with the "pine tree logo" on the gable ends, lava rock foundation and chimney, massive front door with wrought-iron hardware, and knotty-pine paneling inside—is a typical CCC-built guard station. Because of nearby Imnaha Springs that flow from a lava flow and the shade of the big trees, the Imnaha area—its historic guard station and its campground—remains a favorite destination.

ACCESS: Imnaha Guard Station is reached by paved roads (two-lane and single-lane with pull-outs), except for the last 1/4 mile which is gravel road, from either Prospect, about 12 miles away, or Butte Falls, about 18 miles away. From Prospect, turn east off Oregon Highway 62, the Crater Lake Highway, at the Prospect turn-off just south of Prospect Ranger Station, drive through Prospect, turn east on the Prospect-Butte Falls Highway, and bear left onto Forest Road 37 after about 2 or 3 miles. Drive for about 7 miles to the Imnaha Campground sign, and turn left onto the gravel road. Imnaha Campground is a short distance down the road on the right, and Imnaha Guard Station is a little farther along on the left. From Butte Falls, take the Prospect-Butte Falls Highway north a mile or less east of town, drive north on the highway for about 9 miles, and turn right onto Forest Road 34, also called Lodgepole Road. Follow Forest Road 34 about 8 miles, turn left onto Forest Road 37 to the Imnaha Campground sign, and turn right onto the gravel road to the campground and guard station.

LAKE OF THE WOODS RANGER STATION
Winema National Forest, Oregon
(1937)

The former Lake of the Woods Ranger Station, now the Lake of the Woods Visitor and Work Center, is yet another outstanding example of the Pacific Northwest Region's Cascadian Rustic style of Depression-era ranger stations.

Lake of the Woods Ranger Station was built by the Civilian Conservation Corps between 1937 and 1939 in the dense, old-growth, mixed-conifer forest at the northern end of beautiful Lake of the Woods. It served as a Rogue River National Forest district ranger station until 1961, when the Winema National Forest was formed from the Klamath Indian Reservation, that part of the Rogue

Former district ranger's office (above), now a visitor center, and former ranger's residence (below) at historic Lake of the Woods Ranger Station, Winema National Forest, Oregon.

River National Forest containing Lake of the Woods, and lands carved out of the Deschutes and Fremont national forests.

All but one of the Lake of the Woods Ranger Station buildings—the office, two residences, crew house, garage, warehouse, and gas and oil house—are standard Cascadian Rustic wood-frame structures on poured concrete foundations faced with native stone. All are in good condition and, except for new metal roofs, retain their original appearance. The office is easily identified by its location close to the highway and by the words "VISITOR CENTER" below the distinctive "pine tree logo" on its front gable. The barn, a 1 1/2-story peeled-log structure with a wood-shingle gambrel roof, is the compound's architectural exception.

The historic Lake of the Woods Ranger Station office building, its knotty-pine panelled lobby dominated by a massive stone fireplace, houses exhibits depicting early Forest Service ranger and fire fighting activities and Klamath Indian use of native plants. Visitor information is available.

ACCESS: Historic Lake of the Woods Ranger Station is located on the south side of Oregon Highway 140, 33 miles west of Klamath Falls and 43 miles east of Medford, and is open daily during the summer.

PAISLEY RANGER STATION
Fremont National Forest, Oregon
(1937)

The small south-central Oregon town of Paisley is the home of the 1,300,000-acre ZX Ranch, the Paisley High School Broncos, an annual Mosquito Festival, and the Depression-era Paisley Ranger Station.

Still the headquarters of the Paisley Ranger District, the Paisley Ranger Station compound was built between 1937 and 1939 by a work force of Civilian Conservation Corps enrollees and local experienced men under Forest Service supervision. The entire compound—a ranger's residence, office building, barn, warehouse, garage, and gas house—cost Uncle Sam less than ten thousand dollars.

Although a "new" office building was added in 1963 and the CCC-constructed buildings have been modified to accommodate current uses, the compound retains the feel and flavor of a pre-World War II ranger station. The district ranger still lives in the 1938 ranger's residence.

ACCESS: Paisley Ranger Station, on Oregon Highway 31 in Paisley, Oregon, is open during regular business hours. A self-guided tour of the ranger station compound is available at the main office.

GLACIER RANGER STATION
Mt. Baker-Snoqualmie National Forest
(1938)

Adjectives like "quaint" and "picturesque" are used often to describe the historic Glacier Ranger Station office building that, nestled in the forest on the east side of the town of Glacier not many miles south of the Canadian border, is one of the more striking legacies of Depression-era public works programs on America's national forests. Built by the Civilian Conservation Corps and local experienced men in 1938, this native stone and timber building housed the old Glacier Ranger District's headquarters for more than four decades. It remains in service today as the Glacier Public Service Center operated jointly by the Forest Service and National Park Service for Mt. Baker-Snoqualmie National Forest and North Cascades National Park visitors.

Glacier Ranger Station is one of three or four Glacier-area headquarters from which forest rangers operated before World War II. A log cabin that pioneer Ranger John W. Barber—appointed by the Department of the Interior to the North Fork Nooksack District of the Washington Forest Reserve in 1899—was assigned to build at Glacier Creek may or may not have been the first. Gallop Ranger Station, built in 1907 on Gallop Creek in the town of Glacier, was the first Forest Service ranger station from which the North Fork Nooksack District—soon known as the Glacier Ranger District—of

217

Gallop Ranger Station, in the town of Glacier (above, U.S. Forest Service photo), preceded Glacier Ranger Station. Historic Glacier Ranger Station, now Glacier Public Service Center, Mt. Baker-Snoqualmie National Forest, Washington (below).

Landscaping the Glacier Ranger Station grounds in 1938 (above), and Glacier Ranger Station in 1939 (below). U.S. Forest Service photos.

the then Washington National Forest was administered. This station consisted of a combination ranger's office and dwelling, a shop, a warehouse, and a barn from which Ranger Cliff "C.C." McGuire ranged over his 225,860-acre district looking after logging, mining, and other activities of the day. In 1924, the Washington National Forest became the Mt. Baker National Forest. By 1927, the first Glacier Ranger Station—east of Glacier Creek and relocated a couple times during the ensuing decade by highway construction—was district headquarters.

Then, in 1938, the CCC built a new Glacier Ranger Station. Construction took several months, and the biggest part of the job was the masonry. Columnar basalt, quarried during the summer near Heather Meadows at the head of the Mt. Baker Highway, was trucked to the site where a journeyman mason—a local man doing CCC service—did most of the laying. The framing, roofing, and finishing were done by the CCC crew. The result—a sturdy, well-crafted, Alpine Bungalow style building of basalt masonry with wood-frame wings—was a work of art. And the initials "U.S." and "F.S." flanking the word "OFFICE" carved into the squared-timber cross member beneath three large "pine tree logo" cut-outs of its center gable left no doubt about the building's purpose.

The building's basalt masonry is repeated in the curbs and fences on the station's beautifully landscaped grounds. Another feature the CCC contributed to those grounds was a 730-year-old, eight-foot diameter Douglas-fir "history tree" section mounted on a flagstone platform in 1939. A replacement log now occupies that platform. Also still in place, at the southwestern corner of the 1938 office, is the office building of the 1927 ranger station later used as a residence.

By the late 1960s, most of the old Glacier Ranger District's backcountry had been incorporated into the new North Cascades National Park, established in 1968. Eventually, the rest of the district was placed in the Mt. Baker Ranger District headquartered in Sedro-Woolley, and Glacier Ranger Station's job as a district ranger's headquarters was done. But, as the Glacier Public Service Center, it continues to serve throngs of national park and national forest visitors every summer.

ACCESS: Historic Glacier Ranger Station is located 38 miles east of Bellingham and Interstate Highway 5, and just east of Glacier on the south side of Washington Highway 542, the Mt. Baker Scenic Byway. Operated as the Glacier Public Service Center, it is open during the summer season only from 9:00 a.m. to 5:00 p.m. daily.

Ranger boat Chugach, *moored at the U.S. Government Dock in Petersburg, Alaska, in 1989. U.S. Forest Service photo.*

Chapter Seven
ALASKA REGION

Alaska, the largest state, harbors the largest national forests. The sixteen million-acre Tongass National Forest and the five million-acre Chugach National Forest, equal in area to the state of Maine, comprise one-sixth of the state's land area. Spanning southeastern Alaska from the Kenai Peninsula through the Alexander Archipelago, these temperate coastal forests of valuable Sitka spruce and hemlock—Alaska's best—are an extension of the Pacific

Historic Petersburg Ranger Station and Ranger Boat Chugach *are located in Petersburg, Alaska, in the Alaska Region (Region 10).*

223

Northwest rain forest. Over half of this national forest acreage is on islands. On the deeply fjorded mainland, large glaciers abutting coastal waters and steep terrain hamper overland travel. And so, in Alaska, the Forest Service went to sea. Indeed, if the Alaska Region has a distinctive ranger station architectural style, it's the distinctive naval architecture of its ranger boats.

Both historic Forest Service "ranger stations" in the Alaska Region are located in the Kupreanof Island town of Petersburg, headquarters of the Stikine Area of the Tongass National Forest, reached from the mainland by air and ferry. Here is berthed the 1925 ranger boat, *Chugach,* and here is the 1937 former Petersburg Ranger Station. *Chugach,* the last wooden ranger boat in Forest Service ownership, is still in service and has become a symbol of the region. The old Petersburg Ranger Station has become a visitor center. Both reflect unique responses to the challenge of managing Alaska's national forests.

RANGER BOAT CHUGACH
Tongass National Forest, Alaska
(1925)

Ranger boat *Chugach* is the last of the wooden Forest Service ranger boats that, since 1908, have plied the 12,000-mile coastline of the nation's two largest national forests, the Chugach and the Tongass of southern Alaska. She remains in service today as a fully functioning ranger boat.

Known unofficially as the "Green Serge Navy," the ranger boat fleet served as afloat ranger stations that—in the days before adequate aircraft services—provided access to Alaska's dense coastal and island forests. William Weigle, Alaska forest supervisor from 1911 to 1919, explained the difference between the ranger's world in Alaska and in the "lower 48" that made ranger boats necessary this way:

> The motor boat took the place of the saddle and pack horse; hip boots and a rain slicker the place of chaps; and it was much more

essential that a ranger know how to adjust his spark plug than be able to throw a diamond hitch. His steed may do just as much pitching and bucking, but this is prompted not by a spirit of animal perversity but by the spirits of climatic adversity. He guides his steed by means of a wheel instead of reins; feeds it gasoline instead of oats; tethers it at night by means of an anchor in some sheltered cove instead of a picket rope in a mountain meadow; and uses a paint brush in lieu of a curry comb.

And one E.A. Sherman, in a 1921 report to the supervisor of the Chugach and Tongass national forests, extolled the virtues of Alaska's ranger boats in the years just before *Chugach* joined the fleet:

The Alaskan ranger is just as proud of his boat as the Bedouin horseman is of his steed, and the Ranger boats in Alaska are the most distinctive craft sailing the waters of the Alexander Archipelago. . . . They are staunch boats . . . (possessed of) strength, seaworthiness, and a special ability for the particular service expected of them. In case of any trouble or disaster in Southeastern Alaska, . . . the public appeals to the nearest Ranger boat. . . .

Chugach was designed for the Forest Service in 1925 by Puget Sound naval architect L.H. Coolidge, best known for his World War II *Miki*-class tugs, and built that same year at the Lake Union Dry Dock and Machine Works in Seattle for $26,185. About 62 feet long and 14 1/2 feet wide, she displaces about 40 tons and draws about six feet of water. Her below-decks spaces—from stem to stern—include a fo'c'sle with four berths, an engine room with adjoining head and shower, and a trunk cabin with wardroom and galley. Topside, the pilothouse sits over the engine room and contains the engine controls, chart table, navigation and radio gear, and the original wood and brass steering wheel. Two bronze Forest Service shields are mounted port and starboard. Originally painted white with a crimson pilothouse and cabin, *Chugach* received a coat of World War II gray, was dark green in the late 1940s and early 1950s, and has been white ever since. A succession of lifeboats, skiffs, and whalers, a progression of masts rigged with sails and arrayed with ever more modern antennas and lights, and radar recently installed atop the pilothouse have completed the picture over the years.

Since her launching in 1925, *Chugach* has been in continuous service as part of a fleet that, at its peak in the 1920s, numbered eleven ranger boats from which Forest Service personnel carried out myriad tasks. Captained for her first twenty-three years by Erland M. Jacobsen, a Danish-born seaman employed by the Forest Service as both boat operator and forest ranger, she operated out of Cordova, then Chugach National Forest headquarters, providing the vital link between forest administrators and forest users. Indeed, with a complete duplicate set of office records on board, *Chugach* was the principal locus of Chugach National Forest administrative work throughout those years. But her service wasn't limited to routine forest administrative duties such as timber management, special use permit administration, and transporting Civilian Conservation Corps work crews to work camps and Forest Service officials on inspection tours.

As the only federal boat on year-round duty in Prince William Sound, *Chugach* performed frequent search and rescue and other public service missions. In addition to many dramatic rescues at sea, her logbooks tell many tales of transporting medical personnel and supplies as well as emergency food shipments to remote towns and Indian villages; carrying game wardens to distribute hay to feed starving deer; relighting U.S. Lighthouse Service lights and buoys; transporting prohibition agents, other law enforcement officials, personnel of many government agencies, scientists, ministers, and charity workers to perform their duties and ministries; serving as a floating scientific research station; carrying the mail; and transporting U.S. troops during World War II. On occasion, she took Cordova townsfolk to annual picnics, and high school students and their teachers on two-day "sneaks." The first duty assigned to Spencer N. Israelson, second skipper of *Chugach,* was counting every person and family residing in remote areas—outside main towns like Cordova and Whitier—of Prince William Sound for the 1950 U.S. census.

During such a career, *Chugach* was bound to have her share of mishaps. In May, 1948, while Jacobsen and Ranger James Clough were ashore scaling logs on Montague Island, the boat dragged her anchor and went aground. Afloat again, with no apparent serious damage, she was later in danger of sinking in rough seas. An inspection revealed "two holes, five damaged planks, two broken ribs, and a severely damaged keel and keelson" that Jacobsen patched before returning to Cordova, where permanent repairs were made.

Things changed for *Chugach* in the late 1940s and early 1950s. First, at the end of 1949, Jacobsen retired, and a temporary skipper—Harold "Andy" Anderson, the new Chugach National Forest supervisor who had served as a U.S. Navy PT boat commander during World War II and had sailed with Jacobsen as early as 1946—came aboard. Within a few months, Israelson, the "Petersburg fisherman" Anderson had sought for the job, came aboard as full-time boat operator. Then, in 1953, after work on the Chugach National Forest had fallen off considerably, *Chugach* was permanently transferred from Cordova to the Tongass National Forest, which was experiencing a dramatic increase in timber production in response to post-war demands. This was not new territory for *Chugach*. She had been associated with the Tongass on temporary assignments.

But a fierce storm made getting *Chugach* to her new berth in Petersburg an ordeal for Captain Israelson and his wife, Frances. In the midst of the storm, during which the boat was often dwarfed by huge seas, a passing ocean-going tug radioed to ask what a "little white boat" was doing out there. But *Chugach,* heavily built with closely spaced ribs, could take the punishment. After twelve days, she made it to Juneau, normally a three-day trip from Cordova, and later went on to Petersburg.

Chugach's service on the Tongass National Forest reflected the changes in Alaska and in the Forest Service that changed the role of ranger boats. With the advent of large-scale timber management and logging operations in the 1950s, Forest Service work changed from the personalized style of the pre-war years to the more organized style of a large agency in the post-war years. Although *Chugach,* under Israelson and two more able skippers, continued to play a vital role, her work became more routine. Dramatic search and rescue missions, typical of the Cordova years, dropped from her logbooks during her southeastern Alaska years.

Ranger boat Chugach *underway in the Wrangell Narrows, Tongass National Forest, southeastern Alaska, 1989. U.S. Forest Service photo.*

Except for a 1954 to 1958 Juneau timber inventory tour followed by a year or two in Sitka, *Chugach* has called Petersburg home port since her 1953 transfer. And two other Petersburg fishermen, Erling W. Husvik from 1961 to 1967 and Arthur Rosvold from 1967 to 1988, skippered *Chugach* through the years of changing and, eventually, declining use of ranger boats. In the recent past, *Chugach* and other ranger boats have provided transportation and offices afloat for a new breed of Forest Service resource specialists and multi-disciplinary teams engaged in research and other field work on Alaska's national forests. Today, in addition to *Chugach*—which completed a major repair period in Port Townsend, Washington, in 1995, and has returned to Petersburg—only two other ranger boats, *Sitka Ranger* operating out of Sitka and *Tongass Ranger* operating out of Ketchikan, remain in service.

ACCESS: *Still in commission,* Chugach *is moored at the U.S. Government dock on the Petersburg, Alaska, waterfront.*

*Historic Petersburg Ranger Station, Petersburg, Alaska, in 1985.
U.S. Forest Service photo.*

PETERSBURG RANGER STATION
Tongass National Forest, Alaska
(1937)

Another old ranger station was saved from destruction when the community of Petersburg, Alaska, rallied in the late 1980s to preserve part of its heritage and gained a first rate visitor center in the process.

The building that housed the Petersburg Ranger District for over thirty-five years seems to have been justified as more than just a ranger station and to have been made possible by New Deal funds and labor. When, in 1935, a "proposed office building" was being discussed, there seemed "little if any hope of getting a Federal building for this town." But District Ranger J.M. Wyckoff aggressively pursued the project with other officials who needed office space. In 1936, when Regional Forester Charles H. Flory requested funds from the Chief Forester, he noted: "It would be to our advantage to have a building which would also house the Customs

Office, as it could during the Ranger's absence give the public information that they might require. . . ."

And so, with Emergency Relief Act funds, a 50-foot by 100-foot corner lot was purchased in 1936 for $650, and architect W.A. Manley designed a Colonial Revival style, two-story, split-foyer office building and adjoining garage for the site. Labor for site preparation and construction was provided by locally-hired Civilian Conservation Corps members, and the office and garage buildings were completed in the fall of 1937. Each floor· of the 24-foot by 28-foot wood-frame office building provided 672 square feet of office space, and the garage measured 24 feet square. A semi-circular wooden arch marked the main entrance of the otherwise plain, square structure.

From 1937 until 1957, the building housed the Petersburg Ranger District office as well as the U.S. Customs Service office and the Clerk of the Alaska Territorial Court. After 1957, the building was used by the Forest Service alone until, in 1985, all personnel were moved to new quarters and the old office was closed. In a 1987 land exchange with the Forest Service, the City of Petersburg acquired the buildings.

The city considered wrecking the office and garage and turning the site into a parking lot. But many citizens thought the historic office building should be preserved and used. After considerable debate at local coffee shop gatherings and city council meetings, they decided to turn the lower floor into a visitor information center. The combination of the building's central location (just one block from the business district and one block from the museum), the fact it could be modified for wheelchair access, and its historic roots made it an ideal place at which to provide information to people who visit and live in this isolated southeastern Alaska fishing community.

The local chamber of commerce leased the building for a dollar a year, and led the effort to renovate and use the abandoned structure. The Forest Service entered into a partnership agreement with the chamber to assist in renovation and to share in operation of the visitor information center. Local Forest Service archaeologists and landscape architects assisted in designing the renovation to assure the building's historic character was preserved while

230

structural and accessibility modifications were made. Forest Service interpretive specialists helped plan the layout and function of the center. And many Forest Service employees joined other local volunteers in the renovation work. Funds came from a variety of government and private sources. The "new" visitor information center officially opened on May 28, 1992, with many of those who contributed to the effort in proud attendance.

The old Petersburg Ranger Station building now features a fully accessible visitor center on its lower level and an expanded Chamber of Commerce office and community conference room on the upper floor. Visitors are oriented to the community and surrounding forests and waters by exhibits and knowledgeable staff and volunteers. A computer in the center enables visitors to reserve remote Tongass National Forest public recreation cabins.

ACCESS: The old Petersburg Ranger Station building is located at the corner of Fram and First streets in downtown Petersburg, Alaska, reached from the mainland by ferry or aircraft. The visitor center, staffed by Forest Service and Chamber of Commerce folks, is open seven days a week during the summer months.

ACKNOWLEDGMENTS

I didn't realize, until I was too committed to quit, how much work putting *Uncle Sam's Cabins* together would entail. The project could not have succeeded to the extent it has without the help of many sources of inspiration, encouragement, information, and photographs. Most, but not all, are Forest Service sources. All deserve recognition and thanks.

In the Washington Office, thanks to Dr. Terry West, Historian, for useful background information and for informing Forest Service colleagues around the West about the project, and to Evan De-Bloois, Historic Preservation Officer, for information about ranger and guard stations on the National Register of Historic Places.

The introductory chapter about early-day forest rangers and ranger stations reflects myriad sources of information as well as many years of investigation. Primary sources, including early 1900s editions of the *Use Book* that guided early Forest Service rangers on the job, proved invaluable. This chapter is further edified by fair use of well-written words of others, and I gratefully acknowledge permission of the following publishers to quote brief passages from their copyrighted material: Island Press, Washington, D.C., and Covelo, California, for passages from *Breaking New Ground* by Gifford Pinchot, copyright © 1947 by Estate of Gifford Pinchot, renewed © 1974 by Gifford B. Pinchot; The University of Chicago, Chicago, Illinois, for a passage from "USFS 1919: The Ranger, the Cook, and a Hole in the Sky" in *A River Runs Through It* by Norman McClean, copyright © 1976 by The University of Chicago; The Caxon Printers, Ltd., Caldwell, Idaho, for a passage from *The Big Blowup* by Betty Goodwin Spencer, copyright © 1958 by Betty Goodwin Spencer; The University of Nebraska Press, Lincoln, Nebraska, for a passage from *Fire* by George R. Stewart, copyright © 1948 by George R. Stewart; and Simon & Schuster, Inc., New York, for a passage from *English Creek* by Ivan Doig, copyright © 1984 by Ivan Doig.

At the Northern Region headquarters in Missoula, Jud Moore of the public affairs office gave good advice and arranged for Joyce Pritchard of that office to entrust me with a collection of historic photographs from which the pieces on Alta, Bull River, Judith River, Slate Creek, and Ninemile ranger stations benefitted. Cass Cairns, Information Assistant, Bitterroot National Forest, and Linda King, Resource Clerk, West Fork Ranger District, provided information on Alta Ranger Station. A poignant quotation from "The First Ranger Station" by Peyton Moncure in the Sunday, October 4, 1959, edition of the *Salt Lake Tribune Home Magazine,* and a brief quotation from *Whose Woods These Are* by Michael Frome, copyright © 1962 by Michael Frome and published by Doubleday & Company, Inc., New York, add to this piece. Rebecca S. Timmons, Archaeologist, Kootenai National Forest, and Jim Mershon, District Ranger, Cabinet Ranger District, provided Bull River Ranger Station information. Cort Sims, Archaeologist, Idaho Panhandle National Forests, reviewed profiles of Avery Ranger Station prepared with the assistance of Jamie T. Schmidt, Recreation Specialist, Avery Ranger District, St. Joe National Forest, and of Luby Bay Ranger Residence prepared with materials provided by Debbie Wilkins, Recreation Forester, and Marcella Cooper of Priest Lake Ranger District, Kaniksu National Forest. Bonnie Dearing, Public Affairs Officer, Lewis and Clark National Forest, and Larry Timchak, District Ranger, Judith Ranger District, contributed information and photographs for the Judith River Ranger Station entry. Gary Solberg, Slate Creek Ranger Station, showed me around Slate Creek Historical Ranger Station on the Nez Perce National Forest where I took the cover photograph. I am especially indebted to District Ranger Dennis Dailey, Moose Creek Ranger District, Nez Perce National Forest, who was generous with his time when I showed up unexpectedly at his Grangeville office and with source materials, photographs, and comments. Jan Halverson, now-retired Public Affairs Assistant, Flathead National Forest, was generous with documentation on Spotted Bear Ranger Station and loaned me her personal copy of Charlie Shaw's *The Flathead Story* published by the Flathead National Forest in 1964. Carol Hennessey, Recreation Forester, Lochsa Ranger District, Clearwater National Forest, was equally generous with her time, materials, and historical photographs of Lochsa Historical Ranger Station, and volunteers Mr. and Mrs. Chester Fiscus of Potlach, Idaho, cheered my two cold and rainy visits to that most visitable historic ranger station with warm hospitality and hot coffee. Thanks, also, to Diane Converse, Publications Manager, Northwest Interpretive Association, for permission to quote from *Lochsa* by Louis F. Hartig, copyright © 1989 by Pacific Northwest National Parks and Forests Association. And thanks to J. Michael Lunn, Supervisor, Siskiyou National Forest, for sharing his memories of Lochsa's restoration. District Ranger Greg Munther and Laurie Kreis, Pat Perry, and Lynn Sholty of his Ninemile Ranger District staff provided information on and photographs of Lolo National Forest's magnificent Ninemile Remount Depot and Ranger Station, while Bob Hoverson, Pack Train Manager, filled me in on Northern Region Pack Train operations. Jane Reed Benson's *Thirty-two Years in the Mule Business,* published in 1980 by the Forest Service, proved a useful Ninemile source. Cynthia Lane, District Ranger, and Chris Talbert, Receptionist, Selway Ranger District, Nez Perce National Forest, provided information on Fenn Ranger Station and recommended *Major Fenn's Country* by Neal Parsell, published by the Pacific Northwest National Parks and Forests Association, an excellent source. Judy Powers, Business Management Assistant, Darby Ranger District, and Dorothy Goodrich, a volunteer, welcomed me at Bitterroot National Forest's excellent Darby Historical Ranger Station. A. Richard Guth and Stan B. Cohen's *Northern Region,* copyright © 1991 by the authors and published by Pictorial Histories Publishing Company, and Robert D. Baker, *et. al.,* in *The National Forests of the Northern Region,* prepared for the Forest Service by Intaglio, Inc., in 1993, provided useful Region 1 perspective.

I am deeply indebted to Jim Schneck, Architectural Historian at the National Park Service's Midwest Archeological Center, Lincoln, Nebraska, and to Douglas D. Scott, Acting Assistant Regional Director, Anthropology and Archeology, and Dr. Ralph Hartley, Project Archeologist, at that center, for extensive assistance in developing the chapter on Rocky Mountain Region historic ranger stations that, among other administrative structures, the National Park Service is researching for the Forest Service. The profiles of Cayton, Hog Park, Lost Man, Alpine, Turkey Springs,

Michigan Creek, Dolores, and Mesa Lakes ranger stations owe much to this collaboration. Terri Liestman, Archaeologist, Rocky Mountain Region, advised her colleagues of the project. Many thanks, also, to William Touches Deer Puckett, Heritage Program Manager, Shoshone National Forest, for providing extensive historical documentation and oral history transcripts on which the Wapiti Ranger Station piece is based. Gary Osiur, Assistant District Ranger, Rifle Ranger District, White River National Forest, was helpful with the Cayton Ranger Station effort, and Elaine Langstaff, Visitor Information Specialist, loaned historic photographs from Mrs. James G. Cayton's album. Very special thanks to Mr. John Baird of Pagosa Springs, Colorado, for helping me get his Turkey Springs Ranger Station story right. Jim Heid, Archaeologist, Medicine Bow National Forest, provided historical documentation on several historic ranger stations from which I selected Jack Creek Guard Station and Brush Creek Ranger Station, and Dave McKee, West Zone Archaeologist, came through with excellent photographs of both. A neighbor, Milt Griffith, retired Deschutes National Forest deputy supervisor and former district ranger on the old Glade Ranger District, San Juan National Forest, suggested and provided the recent photograph of Dolores Ranger Station. Very special thanks to Sally Crum, North Zone Archaeologist, Grand Mesa, Uncompahgre, and Gunnison National Forests, for her enthusiastic support—including extensive photographic support—and for allowing access to her Collbran Ranger Station files on both the Mesa Lakes and Collbran ranger stations. Allen E. Kane, Heritage Resource Specialist, Pike and San Isabel National Forests, provided valuable information on and photographs of Lake George Ranger Station. A source he led me to, Mr. William O'Shia of Portland, Oregon, shared his CCC experience helping build the station. Jerry Davis, South Park Ranger District, provided a useful 1990 letter from Wendell R. Becton of Gainesville, Georgia, who was district ranger at Lake George when the station was built.

The chapter on Southwestern Region historic ranger stations begins with Hull Tank and Jacob Lake ranger stations for which Teri Cleeland, Historian, Kaibab National Forest, provided information and photographs. Bruce R. Donaldson, Sitgreaves Zone Archaeologist, Apache-Sitgreaves National Forests, provided information on Los Burros Ranger Station that included a fact-packed article by Jo Baeza entitled "Los Burros: A Lot of History in a Little Campground" that appeared in the Thursday, February 19, 1987, issue of the *NavApache Independent*. Urging me on with the project, Dr. Jon Nathan Young, Carson National Forest, provided generous documentary and photographic support on the Old Tres Piedras Administrative Site. Cheryl Oakes, Librarian-Archivist, Forest History Society, Durham, North Carolina, made the historic photograph of Aldo Leopold available, and The University of Wisconsin Press granted permission to quote from *Aldo Leopold: His Life and Work* by Curt Meine, copyright © 1988 The Board of Regents of the University of Wisconsin System. A neighbor, Dick Spray, a retired Southwestern Region recreation staff officer, recommended historic White Creek Ranger Station and put me in touch with Chuck Hill, Los Lunas, New Mexico, the last district ranger to use it as a summer headquarters. Robert H. Schiowitz, Archaeologist, Gila National Forest, was generous with historic documentation that included former Ranger Henry Woodrow's fascinating 1943 memoir of White Creek Ranger Station and the old McKenna Park Ranger District. James A. McDonald, Archaeologist, Coronado National Forest, provided documentation on several historic ranger stations and helped with the Lowell Ranger Station and Cima Park Fire Guard Station profiles. Dr. Ken Kimsey, Museum Curator, Prescott National Forest, proved a great source of information on and photographs of Crown King Ranger Station. William A. Westbury, Archaeologist, Canjilon Ranger District, Carson National Forest, sent information on and photographs of Canjilon Ranger Station. Thanks, finally, to David "A" Gillio, Historian, Southwestern Region, who, with the assistance of Joe Hereford, provided historic photographs of several ranger stations in the region.

Jerry Wylie, who looks after heritage resources and tourism for the Intermountain Region, helped as a source of information on and passer of word about the project to his colleagues. James R. Schoen, Archaeologist, Bridger-Teton National Forest, sent information and photographs of both "Rosie's Cabin"—the first Blackrock Ranger Station—and diminutive Elk Creek Ranger Station. Chip Sibbernsen, Acting District Ranger, Logan Ranger District, Wasatch-

Cache National Forest, enthusiastically supported my Tony Grove Ranger Station effort with documentation and photographs—and a key to the cabin—and Mike Van Horn, Recreation Technician, followed up with additional materials. Steve Matz, Archaeologist, Salmon National Forest, was generous with materials on both Indianola and California Bar ranger stations. Jack E. Bills, Supervisor, Sawtooth National Forest, responded to my request for information on Valley Creek Ranger Station, and Kathleen Coulter, Assistant Archeologist, provided useful access to Esther Yarber's *Stanley-Sawtooth Country* published in 1976 by Publisher's Press, Salt Lake City, Utah. I discovered historic Pole Creek Ranger Station while in Stanley, Idaho, to check out the Valley Creek station. Larry Kingsbury, Archaeologist and Historian, Payette National Forest, came through with information on and photographs of historic Hays Ranger Station and Warren Guard Station. Rick Schuler, who handles recreation, wilderness, and wildlife on the Mountain View Ranger District, Wasatch-Cache National Forest, sent information on and a photograph of Hewinta Guard Station, and broke the news that the district's old Hole in the Rock Guard Station is no more. Clear across the region, Guy W. Pence, District Ranger, Carson Ranger District, Toiyabe National Forest, put me on to Nancy Thornburg, Director, Alpine County Museum, who proved the key source of information on and photographs of Markleeville, California's, series of historic ranger stations. Mark Sayles, Markleeville Guard Station, provided visitor information. Jerry B. Reese, Supervisor, Targhee National Forest, provided Squirrel Meadows Guard Station information and photographs.

Its dry, bureaucratic title aside, "Contextual History of Forest Service Administration Buildings in the Pacific Southwest Region" by Dana E. Supernowicz, Historian, Eldorado National Forest, proved an invaluable introduction to California's historic ranger stations. The chief editor of that study, Linda Marie Lux, Historian, Pacific Southwest Region, encouraged her colleagues to contact me about the project. The entry on West Fork Ranger Station benefitted not only from information and photographs provided by Michael J. McIntyre, Archaeologist, Angeles National Forest, but also from the generous assistance of Glen Owens, Executive Director, Big Santa Anita Historical Society, who, in addition to historic photographs, sent me a copy of John W. Robinson's *The San Gabriels,* published by the Society in 1991, which proved a valuable reference. Dana Supernowicz put me on to the Eldorado National Forest's historic Pyramid Ranger Station, which I'd seen along U.S. Highway 50 some thirty and more years before while driving to and from my Toiyabe National Forest job, and Judy Rood, Archaeologist, Placerville Ranger District, provided directions to its new location and helped me write about it. Cari S. VerPlanck, Heritage Program Manager, Cleveland National Forest, provided information on old Laguna Ranger Station, now called El Prado Cabin. Lynda Martin, Vice President, Laguna Mountain Volunteer Association, provided photographs—including permission to use one of the Association's copyrighted photographs—and additional information, and Gregory S. Greenhoe, Deputy Fire Management Officer, Angeles National Forest, who played the part of Ranger Carl Brenner at the first Living History Weekend at El Prado Cabin in 1986, added color commentary. Carmel Barry-Meisenbach, Historian, Tahoe National Forest, was generous with information on and photographs of three historic ranger stations, and gave me a copy of Richard Markley's interviews with Ranger Frank W. Meggers published by the Tahoe National Forest in 1993 as "Recollections of the Tarweed Kid." Dennis Stevens, Archaeologist, Downieville Ranger District, assisted on Camptonville. The story of the Camptonville forest fire is derived from Montgomery M. Atwater's *The Forest Rangers* published by Macrae Smith Company in 1969. Connie Popelish, Archaeologist, Minarets Ranger District, Sierra National Forest, contributed information on and photographs of three very different historic stations. I enjoyed a great trip to Patterson Ranger Station with Kathy Pitts, Historian, Modoc National Forest, a most enthusiastic source. Kathleen A. Jordan, District Ranger, Monterey Ranger District, Los Padres National Forest, and Andrea Maliarik, Archaeologist, shared Indians Guard Station. Ken Wilson, Heritage Resources Program Manager, Six Rivers National Forest, provided Gasquet Ranger Station documentation. Mark Arnold, Weaverville-Big Bar Zone Archaeologist, Shasta-Trinity National Forests, provided information on and a tour of Weaverville Ranger Station.

I am indebted to many in the Pacific Northwest Region. Susie Graham, Visitor Information Specialist, Hood Canal Ranger District, and Lil Stinson, a volunteer, supplemented my visit to the Olympic National Forest's historic Interrorem Ranger Station with information and photographs. On the Ochoco National Forest, Terry Fifield, Archaeologist, Snow Mountain Ranger District, took me on an October 1993 trip to historic Allison Ranger Station and provided essential information. After his transfer to Alaska, Kyrie Murphy, his successor, followed up with historic photographs and good advice. Jeff LaLande, Archaeologist, Rogue River National Forest, was my source for Star Ranger Station and Imnaha Guard Station, and Barbara Mumblo, Botanist and Historian, Applegate Ranger District, pitched in on Star. Jim Denney, well-known Oregon artist and the Willamette National Forest's guard at historic Fish Lake Ranger Station for many years, introduced me to that historic ranger station and remount depot. Thanks, also, to Randy Dunbar, Recreation Staff Officer, Willamette National Forest, for improving the Fish Lake profile, and to Ron Johnson, Forest Service retiree and Oregon Director of the Forest Fire Lookout Association, for the historic Fish Lake photograph. Research on Depression-era ranger stations in this region was enhanced by "Utterly Visionary and Chimerical: A Federal Response to the Depression," a 1979 Master of Arts thesis by Elizabeth Gail Throop, Historian, Pacific Northwest Region, about CCC construction on Washington and Oregon national forests. Alex Bordeau, Archaeologist, Fort Rock Ranger District, Deschutes National Forest, supplemented my research on Cabin Lake Ranger Station with advice and information. Jan Hollenbeck, Archaeologist, Mt. Baker-Snoqualmie National Forest, provided access to information on Glacier, Monte Cristo, and Silver Creek ranger stations, and Carol Cassity, White River Ranger District, provided a photograph of the Silver Creek cabin better than the one I took. Barbara Kennedy, Acting District Ranger, Bear Springs Ranger District, Mt. Hood National Forest, who told me I'd miss "the cream of the crop" of Region 6's historic ranger stations if I missed Clackamas Lake Historic Ranger Station, inspired the visit that convinced me and provided the information I needed. Pete Martin of that district provided the photograph of Smokey Bear on the 1935 fire truck. A

March 1993 visit to Gold Beach Ranger Station brought assistance from Janet E. Joyer, Archaeologist, Siskiyou National Forest, and Tex Martinek, Archaeologist, Susan Mathison, Public Affairs Officer, and Steve Foster, Maintenance Supervisor, Gold Beach Ranger District. Nancy Rose, District Ranger, Bly Ranger District, Fremont National Forest, and Ed Doremus, Facilities Manager, were sources of information on Bly Ranger Station. In October 1993, Chris Thompson, Archaeologist, Klamath Ranger District, Winema National Forest, drove me up to historic Lake of the Woods Ranger Station and loaned useful documentation. District Ranger Roger King provided a May 1994 tour of Paisley Ranger Station, Fremont National Forest.

For the Alaska Region chapter, Mark McCallum, Archaeologist, Stikine Area, Tongass National Forest, proved a generous source of information on and photographs of ranger boat *Chugach*. The profile of Petersburg Ranger Station benefitted from an article in the October 1992 issue of *America's Great Outdoors* by Richard Estelle, Recreation Staff Officer, Stikine Area, Tongass National Forest.

I am also grateful to most of the above sources for reviewing parts of this work for accuracy. Residual errors, if any, are mine alone. Not among my possible errors are two spellings of one word—"archeologist" and "archaeologist—both used by U.S. Government agencies.

I am especially grateful to a friend and colleague, Viviane Simon-Brown, an educator and articulate editor—and a former Forest Service fire lookout, for editing parts of the manuscript.

Thanks, also, to those who put me up during research trips: Paul and Arlene Alvord of Logan, Utah; Ken and Judy Crandall of North Bend, Washington; Helen Johnson of Placerville, California; and my brother, Dave Joslin, in Dillon, Colorado.

Finally, many thanks to Tom Healy and his staff at Maverick Publications, Inc.—especially Bridget Wise and Patricia Hammer—for helping me realize my vision for this book.

INDEX

A

B

D

E

F

G

Gordon, Granville "Granny" 18
Gordon, Pauline 18
Graham, Joe 201
Grand Canyon Forest Reserve 93
Grand Canyon National Game Preserve 98
Grand Canyon National Park 93 - 94
Grand Mesa National Forest 78 - 80, 85 - 88
Greeley, William B. 14
Grosbeck, Ed 119
Grover Hot Springs State Park 140

H

Hall, C.C. 193
Hall, L. Glenn 168
Hand, Ralph 38 - 39
Hanson, Fred 184
Harriman, W.O. 197
Harrison, President Benjamin 57, 59, 152
Hartig, Louis F. 39, 42
Hartman, Frank 27, 29
Hartman, Virgil "Slick" 29
Hayden National Forest 69 - 70
Hayes, Charles B. 135 - 136
Hays Ranger Station 135 - 136
Hays, Bill 142

Hewinta Guard Station 138 - 139
Higby, Reed 70
Hill, Chuck 109 - 110
Hog Park Guard Station 69 - 70
Holmlund, Victor "Big Vic" 34 - 35
Horton, Lynn 174
Horton, William H. 133
Hot Springs Ranger Station 140, 142
Huffman, Edgar P. 144
Hughes, Jack 65
Hull Cabin Historic District 92
Hull Tank Ranger Station 92 - 95
Hull, Phillip 92
Hull, William 92
Husvik, Erling W. 228

I

Idaho National Forest 135
Idaho Panhandle National Forests 13, 51
Imnaha Guard Station 212 - 213
Indianola Ranger Station 128 - 129
Indians Guard Station 170
Interrorem Ranger Station 184 - 187
Israelson, Spencer N. 226 - 227

M

Mace, Will   97 - 98
Madsen, Ed   180
Maher, Edward J.   10
Major, Merritt B.   184
Malakoff Diggins State
   Historic Park   163
Manley, W.A.   230
Markle, Kathleen   144
Markle, Merle   144
Markleeville Ranger Station
   140 - 144
Marshall, Ed   205
Marshall, Robert   37
Mather Memorial Parkway
   199
Mattison, Tom   39
Maule, William M.   142
McCracken, C.L.   131
McCullough, Nevan   200
McGuire, Cliff "C.C."   220
Medicine Bow National
   Forest   70, 74, 76, 89
Meggers, Frank W.   173 -
   174
Mesa Lakes Ranger Station
   85 - 88
Michigan Creek Ranger
   Station   77 - 78
Mimbres Ranger Station   111
Mitchell, J. Roy   197
Modoc National Forest   168
Mono National Forest   140,
   142
Monte Cristo Ranger Station
   209 - 210

Montezuma National Forest
   67, 80
Moose Creek Ranger Station
   14, 30 - 34
Morey, George   206, 208
Morris, Leonard   177
Mount Olympus National
   Monument   184
Mount Rainier National Park
   199
Mountain View Ranger
   Station   139
Mt. Baker National Forest
   209, 220
Mt. Baker Scenic Byway
   221
Mt. Baker-Snoqualmie
   National Forest   199, 209,
   217
Mt. Hood National Forest
   201 - 202
Mt. Shasta Ranger Station
   167
Myers, Thomas G.   22, 24

N

Nebraska National Forest   57
Nelson, Hans   20
Nelson, Jesse W.   69
Newcomb, Louie   153
Nez Perce National Forest
   14, 27 - 34, 52 - 55
Nielson, Adolph   171 - 172
Ninemile Ranger Station
   44, 47 - 50

R

S

Also by Les Joslin. . .

TOIYABE PATROL

Five U.S. Forest Service Summers

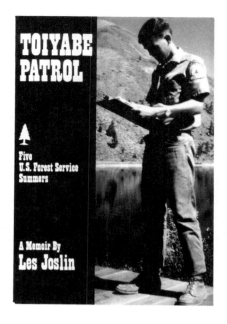

Les Joslin went to work for the U.S. Forest Service on the Bridgeport Ranger District of the Toiyabe National Forest in the summer of 1962.

After a freshman season on the district's fire crew, he was promoted and appointed district fire prevention guard—a seasonal position he held through the summer of 1966.

Hundreds, perhaps thousands, of seasonal U.S. Forest Service employees have served as fire prevention guards on America's national forests. What makes Les Joslin different is not the experience and thoughts related in this book, but the fact that he has chosen to share them.

Toiyabe Patrol is a book everyone who works—or who worked or wants to work—in the Forest Service will appreciate and enjoy.